Winners of the Baldrige Award 1988–1991

1991 Winners

Marlow Industries, Inc.
Public Relations, 10451 Vista Park Road,
Dallas, Texas 75238
(214) 340-4900 Fax: (214) 341-5212

Solectron Corporation
Public Relations, 2001 Fortune Drive,
San Jose, California 95131
(408) 957-7032 Fax: (408) 263-4150

Zytec Corporation
Public Relations, 7575 Market Place Drive,
Eden Prairie, Minnesota 55344
(612) 941-1100 Fax: (612) 829-1837

1990 Winners

Cadillac Motor Car Division
Public Affairs, 2860 Clark Street,
Detroit, Michigan 48232
(313) 554-5065 Fax: (313) 554-5074

IBM Rochester
Public Affairs, Department 70C, Building 107-1,
Highway 52 & 27th Street N.W.,
Rochester, Minnesota 55901
(507) 253-2323 Fax: (507) 253-7175

Federal Express Corporation
Public Affairs, P.O. Box 727,
Memphis, Tennessee 38194
(901) 395-3460 Fax: (901) 346-1013

Wallace Company, Inc.
Public Affairs, P.O. Box 2597,
Houston, Texas 77252-2597
(713) 685-4670 Fax: (713) 672-5848

1989 Winners

Milliken & Company
Public Affairs, P.O. Box 1926, MS 265,
Spartanburg, South Carolina 29304
(803) 573-2546 Fax: (803) 573-2100

Xerox Business Products and Systems
Public Relations, P.O. Box 1600,
Stamford, Connecticut 06904
(203) 968-3378 Fax: (203) 968-4312

1988 Winners

Globe Metallurgical, Inc.
Public Relations, 6450 Rockside Woods
South, Suite 390,
Cleveland, Ohio 44131
(216) 328-0145 Fax: (216) 328-9416

Motorola, Inc.
Media Relations, 1303 East Algonquin Road,
Schaumburg, Illinois 60196
(708) 576-5304 Fax: (708) 576-7653

Westinghouse Commercial Nuclear Fuel Division
Public Relations, Westinghouse Building,
Gateway Center,
Pittsburgh, Pennsylvania 15222
(412) 642-3373 Fax: (412) 642-4985

For information about the Malcolm Baldrige National Quality Award competition, or for a free copy of the guidelines, write or call:

Malcolm Baldrige National Quality Award Office
U.S. Department of Commerce
National Institute of Standards and Technology
Gaithersburg, Maryland 20899
(301) 975-2036 Fax: (301) 948-3716

The Road
to the Baldrige Award

The Electronic Business Series

The Road to the Baldrige Award

~ Quest for Total Quality

Robert Haavind
and the Editors of *Electronic Business*

Butterworth-Heinemann
Boston London Oxford Singapore Sydney Toronto Wellington

Library of Congress Cataloging-in-Publication Data

Haavind, Robert
    The road to the Baldrige Award: quest for total quality / by Robert Haavind and the editors of Electronic business.
       p. cm. — (The Electronic business series)
    Includes bibliographical references and index.
    ISBN 0-7506-9262-6 (case bound)
    1. Malcolm Baldrige National Quality Award.   2. Total quality management—United States. I. Electronic business. II. Title.
III. Series.
HD62.15.H3   1992
658.5'62—dc20   91-45957
                                         CIP

British Library Cataloguing in Publication Data

Haavind, Robert
    The road to the Baldrige Award: quest for total quality. — (The electronic business series)
    I. Title. II. Series.
    658.562

    ISBN 0-7506-9262-6

Butterworth-Heinemann
80 Montvale Avenue
Stoneham, MA 02180

10 9 8 7 6 5 4 3 2 1

Printed in the United States of America

Contents

Vice President Dan Quayle (*end left*) and former Secretary of Commerce Robert Mosbacher (*end right*), with Baldrige Award winners (*left to right*) Winston Chen, Chairman and Chief Executive Officer, Solectron Corporation, San Jose, California; Ronald Schmidt, Chairman, President, and Chief Executive Officer, Zytec Corporation, Eden Prairie, Minnesota; Raymond Marlow, President and Chief Executive Officer, Marlow Industries, Dallas, Texas.

Foreword

Robert A. Mosbacher
Former Secretary of Commerce, Washington, D.C.

We in the U.S. Department of Commerce take seriously the challenge laid down by President Bush to do all in our power to help make the American economy the world's strongest and most prosperous. Central to our effort is the all-important need to act in concert with American business to ensure that U.S. goods and services are competitive and of the highest quality.

The Malcolm Baldrige National Quality Award represents one approach for reaching this goal. This award offers a unique mechanism for cooperation between business leaders and quality experts throughout the private sector and government.

The award criteria detail the requirements for achieving world-class quality. In fact, the success of the Malcolm Baldrige National Quality Award has demonstrated that government and industry, working together, can foster excellence by focusing attention on the importance of the highest quality in U.S. products and services and on the methods of achieving such a level of quality.

There are several indications that the Malcolm Baldrige National Quality Award is successfully raising the level and awareness of total quality management. One indication is that numerous companies have adopted its criteria for self-assessment and training. A second is the growing number of conferences and technical meetings that include sessions on the Baldrige criteria. A third is the proliferation of both state and local recognition programs modeled after the award. A fourth is the additional interest the award is getting from people in areas outside of industry, including education.

There is evidence that American industries are putting quality-improvement strategies to use. Manufacturers are working more closely with suppliers of components for their products, training them in quality management methods, and bringing them into the design phase of new products. Companies are also working to get new products more quickly from the design stage through development and testing to the marketplace.

We also see awareness by service organizations growing very rapidly. Most of the previous winners, although from manufacturing sectors, have large service components. These winners are being asked more and more to speak to top service managers about quality. Many of the lessons learned are transferable across sectors.

We are confident that the Baldrige program will make a significant impact on international trade, but participation in the program must grow to achieve the full benefits in this sector of the economy.

The number of applicants for the Baldrige Award certainly is expected to grow. But more important, we expect the number of companies using the criteria in their own operations and in conjunction with suppliers and customers to expand greatly. I encourage all companies interested in quality to request the application guidelines and use them in guiding their way along the road to world class competition.

The U.S. Commerce Department administers the Malcolm Baldrige National Quality Award program through the National Institute of Standards and Technology.

Foreword

Curt W. Reimann, Ph.D.

In its early years, the Malcolm Baldrige National Quality Award has come to symbolize the resurgence of quality in the United States. The acceptance of the award's tough quality standards underscores the notion that the Baldrige Award is much more than a contest. It is a commitment to excellence. It is an acknowledgment of America's awareness of quality as a competitive tool, and it is a demonstration of the ability of U.S. companies to work together in the national interest.

As further evidence of the impact of the award, some 20 states are now working to develop regional quality programs that are based on the national program. The values represented by Baldrige are carrying over to federal and local government agencies, trade groups, schools, and healthcare centers, weaving together a cooperative program between the public and private sectors on an unprecedented scale.

How can we interpret this phenomenon? One explanation is that the Baldrige Award is a participative, nonprescriptive program rather than a forced march to quality. Applications are evaluated by peer review. The examiners are chosen, primarily from the private sector, on the basis of their personal experience in quality management. Moreover, the Baldrige criteria provide a definition of total quality while helping a company generate evidence of progress. From that perspective, the real goal isn't winning the Baldrige Award but achieving national harmony in the pursuit of excellence. Even if a company does not enter the competition, the criteria may thus be the key to survival in today's increasingly competitive global markets.

Baldrige winners are serving as role models for U.S. industry. Through their investments in human resource development, their dynamic leadership, their articulation of the requirements for competing in the 1990s and beyond, and their unceasing quest of ever-higher quality goals, these companies point the way for all U.S. organizations. By encouraging their suppliers to reach the same goals, the winners have stressed the importance of supplier reliability in achieving the quality needed to compete internationally.

The growing demand for quality is not limited to manufacturing. Customers get just as angry over a billing error as they do over a faulty product. As a result, companies are paying more attention to service and to strengthening communication between customers and suppliers. The Baldrige criteria come into play once again by infusing customer-related quality into the entire internal system and by allowing us to see ourselves through our customers' eyes.

Despite this exciting and encouraging beginning, much remains to be done. The current focus on quality standards in Europe and the continued pursuit of quality in the Pacific Rim suggest that quality leadership will be a rapidly moving target. American institutions need to collaborate in order to ensure the continued growth of the U.S. economy.

Quality-improvement processes must be integrated through team efforts at all levels of involvement. Teachers need encouragement to find new ways to teach, test, and evaluate their students. Students must become well-versed in the procedures that will raise the caliber of their contributions. Pacesetters of all kinds should exhibit the curiosity and enthusiasm that is so desperately needed throughout our educational structure.

The award program is deeply indebted to the many experts who have served as judges, examiners, and overseers and who helped to create and reinforce the program's high standards. In just a single year, these participants and previous award winners will make more than 3,000 formal presentations about the award program, bringing their insight and experience to perhaps 100,000 organizations in business, government, healthcare, and education.

Curt W. Reimann is Director of the Malcolm Baldrige National Quality Award program at the National Institute of Standards and Technology in Gaithersburg, Maryland.

Preface

The Road to the Baldrige Award is a useful resource for any organization striving to achieve excellence in all its operations. It tells the stories of scores of American companies dedicated to total quality through continuous improvement. The companies range across the business spectrum, from the frontiers of high technology to such basic sectors as materials processing, distribution, and package delivery. Some are corporate giants with household names (IBM, Federal Express, Xerox, Motorola, Texas Instruments). Others are less familiar midsize or smaller firms (Perkin-Elmer, Globe Metallurgical, Zytec) or even start-ups (Next Computer).

Because of the scope and variety of the companies profiled, the cross-fertilization of ideas the book provides should prove valuable to members of any organization dedicated to instilling total quality, whether well along in the process or just beginning the journey.

Achieving any major change in corporate culture is no easy task. It becomes particularly difficult, however, when driven by precepts as pervasive as *total quality* and *continuous improvement*. Slogans and management directives built around such lofty ideas will do little to budge traditional mind-sets and patterns of working. Workers who have learned for years that anyone slowing production will probably be fired will be highly skeptical when told that they are being empowered to stop the line to prevent quality problems. Training can help, but it too may be interpreted more as a theoretical exercise than a new way to do one's daily work.

Concepts need to be brought to life. Real examples are essential to demonstrate how total quality methods can be applied at every level in every type of business and to confirm that by following these methods it is indeed feasible to reach extremely tough goals. Doubters also need evidence to verify the tremendous benefits to the organization and its employees that can result from such all-hands efforts.

The wide range of case histories profiled here can help meet such needs. Already, many companies are exchanging visits and ideas with other firms that have set out on the road to total quality. This book serves to extend this idea-sharing process. The need is growing as the quest for total quality extends throughout the American business and industrial community, into smaller firms, subsidiary services, and even public sectors such as education, healthcare, and government.

What's fueling the drive toward total quality now sweeping across the United States? Powerful global competition was a major factor in triggering the movement. Markets once dominated by U.S. firms, both domestically and abroad, have increasingly been challenged or even taken over by foreign firms. Some of the decline can be attributed to lower prices, due in part to cheap labor or less stringent regulation in other nations. Frequently, however, the success of foreign competitors has been due to superior quality in product performance and reliability. In addition, many of these strong overseas companies have proven more agile in adapting to rapidly changing market needs and emerging opportunities.

To meet the global challenge, American corporations have had to reexamine the quality of their goods and services. Careful study of how these tough foreign competitors, particularly those from Japan, are achieving such high quality levels reveals that much more than just final output is involved. The best competitors strive for continuous improvement in *all* their operations and at every level. Although this process is led by management, it depends heavily on a steady stream of ideas contributed by a well-trained, highly committed, empowered work force.

Furthermore, these powerful competitors were found to have achieved tremendous gains in performance without costly investments and new layers of bureaucracy. Instead their costs were reduced, sometimes dramatically, because the high quality at each stage of their operations eliminated the need for extensive testing and rework. Their organizations were flatter than those in most American corporations, with fewer levels and grades, because a flexible work force had been trained in multiple skills. Workers were also challenged to participate in continuous improvement efforts.

The highly successful organizational approach taken by these companies to achieve excellence in everything they do has been termed a *total quality culture* based on *total quality management* (TQM) principles.

How could the United States meet the challenge of these world-class competitors? Some companies, such as Motorola and Xerox, began to adopt many of the methods of total quality on their own several years ago. Numerous studies within companies and across industries showed that as much as 20% or more of total revenues were being wasted on faulty quality (reworking, scrap, field repairs, and so on). It seemed apparent to many executives and policymakers, however, that a much more widespread effort would be required if U.S. industry as a whole was to keep up, or in many cases to catch up, with the total quality achievements of a growing number of foreign competitors. They observed that in Japan a national program of Deming Prizes, named for the American quality guru W. Edwards Deming, had served as a focal point for total quality efforts for nearly four decades. Why not organize a similar program in the United States to instill a total quality culture in American organizations?

The result is the Malcolm Baldrige National Quality Award program. In 1988 the first Baldrige winners were personally congratulated by President Ronald Reagan, and a similar ceremony with President George Bush congratulating the winners has been held each year since. The prestige and influence of the Baldrige criteria, a detailed set of measurement tools against which an organization's excellence can be judged, has grown steadily as the program has become more widely known throughout the American business community.

Among the companies whose total quality efforts are described in this book are Baldrige winners in all award categories: manufacturers, service companies, and small businesses (500 or fewer employees). Other companies have been strong contenders in past Baldrige competitions while still others are laying the groundwork for future entries.

Pioneers in the total quality movement in the United States tended to be large multinational corporations, mostly in the intensely competitive high tech sector, and especially those with subsidiaries or other operations in Japan. These corporate giants, recognizing that their own quality depends on the performance of their suppliers, have been pressuring all those they do business with to adopt total quality methods as well.

Thus, demands are intensifying for *all* U.S. companies to become world-class competitors. Pushing quality to higher and higher levels and cutting cycle times for everything from developing products to making deliveries are forcing close, cooperative partnerships through tiers of suppliers, starting with the giant corporations and reaching to the smallest enterprises.

The United States is not like Japan or any other country that is trying to instill a total quality culture in its business enterprises. Because of our tradition of freedom, Americans, no matter what their ethnic or cultural heritage, tend to be individualists. Homilies and philosophical musings don't easily move managers and workers in the United States. Wherever they live or come from, a U.S. audience tends to have the skeptical response of the proverbial Midwesterner: "I'm from Missouri, you'll have to show me!" This attitude makes it tough for any organization to gain wide support for as pervasive and demanding an agenda as that required by the transformation to a total quality culture. Making it even tougher is an air of cynicism among middle managers and the work force toward any sweeping new programs launched by top management. This all-too-typical attitude stems from a succession of fads in management style and organizational realignment that have swept across the American business scene in recent decades. These approaches were primarily based on top-down, strictly hierarchical methods. They tended to be short-lived, sometimes terminating when imposed goals were reached, but more commonly fading away as management lost interest and latched onto the next fad.

This book helps meet this challenge by showing in detail how all kinds of American companies have launched successful total quality programs, in many cases turning initial skepticism and apathy into enthusiastic support throughout the organization. Although chief executives must be deeply involved in the process, their involvement must be based on a long-term commitment to continuous improvement by an empowered work force rather than the traditional top-down approach. Many of the programs described here struggled into existence, with missteps and setbacks that had to be overcome. Any organization setting out on the quest for total quality will find in these stories practical models of how others like them have succeeded in spite of adversity. The results have often been dramatic improvements, not just in the firm's products or services, but also in its financial performance. Some companies, like Wallace Co., Inc., a Texas pipe distributor, which won a Baldrige Award in 1990, claim that their efforts to meet the stringent criteria set up for the competition helped them not just to survive, but to thrive, when failure and decline had seemed inevitable.

Keeping the total quality flame alive can also be a problem once the most obvious causes of defects have been identified and corrected. In the past, "fad" programs would die out whether or not initial goals were achieved. But continuous improvement is a never-ending process. Each time targets are reached, even though they once may have seemed to be nearly impossible, the bar must be raised still again to new heights. That's because no matter how much better an organization's performance may have become, competitors elsewhere will continue striving to match and eventually to top it. Thus, no organization can remain world class by basking in past glory.

This book should serve as a rich source of fresh concepts and techniques for continually renewing efforts to boost quality even further, to streamline operations, and to better satisfy customers. The great variety of approaches to total quality illustrated reflects the innovativeness that comes from encouraging both individual and team-based initiative.

Winston Chen, the outspoken founder of Solectron Corporation, an electronics contract manufacturing firm whose achievements are cited in Chapter 10, epitomizes the unique vision of total quality developing in the United States. He says Solectron's methods blend the best ideas from Japanese quality leaders with the innovativeness of American workers. Yet, a tour of the manufacturing facility was conducted by an Indian factory manager, and the workers on the line were mostly from Asia, Mexico, and the Caribbean. All these employees, in spite of the diversity of their heritage and cultural backgrounds, were *Americans*. They undoubtedly all hoped they would be able to share in the freedom of opportunity and expression promised to those coming to the United States. The total quality culture was giving each of them an opportunity to contribute in his or her own way to the success of the company he or she had joined as well as his or her adopted country.

Tapping such diversity, and the innovative spirit fostered by a tradition of opportunity, promises to carry total quality to new heights in the United States. To illustrate how this is already beginning to happen, this book is divided into three parts.

Part I, "The Baldrige Award Program," describes the Malcolm Baldrige National Quality Award program. It discusses the reasons why a national quest for total quality is essential if American companies are to be world-class competitors and explores views on the program by Baldrige examiners and the program's director. The criteria developed for the Baldrige program have become widely accepted as a blueprint for achieving world-class status in total quality. In moving toward a total quality culture, it is important to have a uniform template against which progress can be measured, and the Baldrige criteria have become widely accepted as a broad gauge for such evaluations. Information on how to obtain the application guidelines containing the criteria is presented with the list of Winners of the Baldrige Award (1988-1991) and on page 15.

Part II, "Total Quality in Business and Industry," describes specific quality programs, including those of 12 Baldrige Award winners (1988 through 1991) and of many other companies, large and small, hoping to win a coveted award in the future. A great strength of the Baldrige Award criteria is that they are aimed primarily at measuring results rather than the methods employed to achieve them. Because of this nonprescriptive approach, every company profiled has implemented total quality concepts within its own corporate culture. Thus, different lessons can be drawn from each firm's unique experiences. To aid the reader, separate chapters are devoted to specific business sectors—manufacturing, services, small companies, electronics, computers, software, and military contractors. Note that many of the total quality approaches devised in each of these sectors are easily transferable to other types of businesses as well.

Part III, "The Total Quality Approach," describes the emphasis in a total quality culture on such factors as customer satisfaction and education and presents the views of recognized "gurus" of quality. It concludes with a list of resources for further study, a glossary, and an index.

Acknowledgments

This book is the work of many writers and editors dedicated to reporting on the push toward total quality in the United States. Special recognition is due to *Electronic Business* magazine, which published many articles that were adapted for use in the book. The support of the magazine's publisher, Frank J. Burge, and very able former editor, Jeffrey Bairstow, along with his talented editorial staff, were essential to making this project possible.

The book also would not have been feasible without the enthusiastic sharing of information about specific company programs by scores of executives and professionals across the business and industrial spectrum. In addition, the support of Secretary of Commerce Robert A. Mosbacher and Curt W. Reimann, Director for Quality Programs at the National Institute of Standards and Technology, who heads the Baldrige Award program, were instrumental.

The work of several writers was adapted for the book. H. Garrett DeYoung reported on Xerox's outstanding quality efforts and made other important contributions. Bruce Rayner contributed in such areas as education and military quality programs. Rick Whiting helped in describing the overall Baldrige Award program and contributed additional material, including reporting on programs at companies such as Digital Equipment Corporation. Valerie Rice wrote about quality at many smaller firms, especially in the Silicon Valley area. Dwight Davis and Tony Greene wrote about efforts in the software industry.

Other writers making contributions include Elizabeth Baatz, Barbara N. Berkman, Stuart M. Dambrot, Barbara Jorgensen, John Kerr, Giuseppe Labianca, John MacCreadie, Pam Nazaruk, and Teri Sprackland.

Helpful suggestions on improving the manuscript were contributed by Richard P. Schroeder, formerly vice president, corporate quality assurance at Codex Corporation, a division of Motorola, Inc., and currently vice president, quality/time based management, industry segment, Asea Brown Boveri, Inc., Stamford, Connecticut; E. James Tew, a senior Baldrige examiner and manager of quality assurance operations, Texas Instruments, Inc., Defense Systems & Electronics Group; and Jeffrey Bairstow, former editor of *Electronic Business* magazine.

Although credit is due to all of them, the responsibility for the final manuscript remains with the author/editor, who adapted the articles for use in book form and wrote many sections, including those about quality efforts at Motorola, IBM-Rochester, Federal Express, Hewlett-Packard, Solectron, Zytec, Perkin-Elmer, Control Data, and Globe Metallurgical, as well as introductory material and the glossary.

Part I ~
The Baldrige
Award Program

Chapter 1 ~

A New View of Quality

Global competitiveness has become the watchword of American business and industry, and for good reason. Over the decade of the 1980s, U.S. manufacturing productivity growth slowed dramatically while our trading partners made strong gains. While economists puzzled over possible causes for the relative decline, its impact belted the nation right in its midsection. The bad news started with the so-called Rust Belt industries such as automobiles, steel, and machine tools—heartland business sectors that had long been bulwarks of the American economy.

Thousands of plants shut down, and, hundreds of thousands of workers were laid off, an economic slide chronicled day by day and week by week in the media, from daily newspapers and television to magazines and business journals. Eventually, even the newer high-tech industries proved not to be immune. Whatever happened to consumer electronics, once a thriving U.S. industry with dozens of competitors? Or dynamic memory chips, the highest-volume microelectronic devices of all? These ubiquitous digital microchips had once come exclusively from the United States, but by decade's end Japanese manufacturers had captured well over 90% of the global market. Industry analysts feared that computers and microprocessors, the critical components in personal computers and many other electronic systems, might well be the next to go. A stream of government reports expressed increasing concern that American industry was losing its edge in a list of technologies critical to the defense preparedness of the United States.

Economic reports showed that the productivity slowdown was accompanied by a steep rise in the U.S. trade deficit, particularly with Japan and other Pacific Rim nations. Congressional offices were flooded with demands for trade protection against foreign invaders who were "obviously grabbing U.S. markets by selling below their production costs."

But the story didn't end there. During the mid- and late-80s, more penetrating articles, books, government and think-tank studies, even television specials, revealed another side of the story: U.S. manufacturing industries were indeed in decline, and they had much to learn from their fast-rising global competitors, particularly in Japan. Yes, foreign companies did have more support from their governments, and there was good reason to push for more open markets around the world while curbing outright "dumping" abuses in U.S. markets. In addition, however, many offshore factories appeared to be using more modern machinery and processes to achieve higher productivity while making adept use of automation.

Most dramatic of all was the fantastically high quality of the goods some of these foreign factories produced. Executives from U.S. companies such as Xerox and Motorola toured Japanese facilities and found defect levels 500 to 1,000 times lower than those in the average U.S. electronics plant, according to "Made in America," a book that summarized a study of U.S. industries and their international competition by the MIT Commission on Industrial Productivity. This superb quality was being achieved on the production line rather than through extensive testing and rework. In fact, these executives found that most Japanese plants had no final inspection and rework areas at all!

~ Japan's Factories: Living Proof That Quality Pays

The Japanese results amazed the American executives because they seemed to defy traditional management lore. These seemingly incredible quality levels weren't the result of costly systems and extra workers. Quite the contrary, manufacturers with the highest quality levels also tended to have the *lowest* production costs. The Japanese producers were reaching these extremely high quality levels through a wide array of management and organizational techniques, many of them espoused by U.S. consultants who had found little support for their theories in the United States. Other concepts had been developed or refined by Japanese specialists with names virtually unknown in the United States—Ishikawa, Taguchi, and Shingo. Manufacturers like Toyota in automobiles and Matsushita in consumer electronics carried far less inventory and needed much less space for production. Their production lines tended to be more flexible, and they had much better cooperation from their suppliers.

Another clear effect of concentrating on quality was the impact on customers. Brand names such as Nikon, Sony, and Honda were achieving the status gained by such outstanding products as Hasselblad cameras and Rolex watches in an earlier era. Already, Hewlett-Packard had gotten the message across to the U.S. semiconductor industry: The quality level of microchips from Japan was far better than that of U.S. integrated circuits. Buyers were flocking to high-quality, lower cost imported products, from memory chips to automobiles—building market share for foreign producers and thus reducing their per-unit costs even further.

Recognition that quality pays, rather than costs, gradually dawned on many once-skeptical U.S. executives. When they saw that effective quality programs could bolster the bottom line, even the most hard-nosed budget crunchers began to get the message.

In fact, a revolution toward higher quality production was underway in many U.S. companies well before a few industrialists and Congressional leaders started thinking about developing a national program to boost quality in the United States. For decades, quality specialists had been trying to gain the attention of top management in the United States, hoping that some of the continuous improvement methods that had proved so successful in Japan would be adopted by American firms. In the 1970s and early 1980s, executives at some companies began to listen to the total quality message, although initial efforts were superficial, more often lip service and public relations than a real attempt to transform corporate culture.

By the mid- to late-80s, however, a small but steadily growing cadre of U.S.

~ FLAWS IN U.S. MANAGEMENT STYLE?

What's wrong with +~ "

stvle i~ ~~~~~~~~~

i
- V   ...anufacture competitive.
  low end.
- Workers are paid to do, not to think.
- The job of senior management is strategy,
  not operations or implementation.
- The key disciplines from which to draw
  senior management are finance and
  marketing.
- Success is good, failure is bad.
- If it ain't broke, don't fix it.

Don P. Clausing, an MIT professor and
ıber of the MIT Commission on Industrial
ıctivity, suggested that American companies
to find suboptimum solutions because of
ecialization and segmentation within their
ations. Departmental managers often
ɔ with approaches pertinent to their own
:s, while protecting their turf from incur-
other functions. He contrasts this to
industry, where solutions are usually
through team consensus, with teams
nembers actually involved in the work.
sult, Clausing suggests, is that in the
s the quest has been for elegant solu-
ɔblems using definitions that are
ɔbsolete. In Japan, managers seek
tions, using definitions that are
evant.

llustrates this view by contrasting
to inventory storage and retrieval
ıtries. The trend in the United
oward highly automated, com-
high-bay storage and retrieval
ems provide an elegant materi-
lution to a local objective. The
ıas been toward just-in-time (JIT)
in which arrangements are made
ɹliver the right materials and parts
ıre required for assembly. This
ıtes inventory and meets a much
bjective.

perpetuates old objectives,
ıust...   while cross-functional thinking leads to better corporatewide solutions.

companies began serious efforts to upgrade quality to match the world's best, especially in those industries that must compete in the global arena, such as electronics and
computers. It was plain, though, that the bulk of the American business world had not
yet gotten the message. Many thousands of U.S. companies, although keenly aware of
and deeply concerned about growing foreign competition, had little sense of what to
do to match or beat the growing threat to their survival.

Those few that more carefully investigated the success of Japanese and other top global competitors found that their rise in the marketplace was based on far more than simply making products with fewer defects. This was only one result of a much broader concept of quality than was then common in the United States and Europe. These tough challengers had developed a new view of quality, a total quality culture that encompassed everything from how management formulated strategy to how extensively workers were permitted to contribute to defining their own jobs.

The new corporate culture was based on continuous improvement toward ever tougher goals. This called for an organizational structure that encouraged, in fact demanded, a constant flow of new ideas on how to do all kinds of things better. There was no hope that top management on its own could conceive of and introduce such a stream of innovations at every level of the organization. It was essential to enlist the enthusiastic participation of all employees and to scout for better ways to do things from competitors, other industries, and even from around the globe. An important start toward finding such models is a process called benchmarking, in which a company compares its own results to those of competitors in a wide range of activities as diverse as product reliability, costs for processing orders, and customer satisfaction surveys.

To put the current U.S. quality dilemma in perspective, consider that Japan faced a much more severe problem when it launched its own national quality push under the banner of the Deming Prize in 1951. Organized by the Japanese Union of Scientists and Engineers (JUSE), the Japanese program was named for U.S. quality guru W. Edwards Deming. In the early 1950s, the rest of the world viewed Japanese products as junk, and often they were—flimsy gadgets made out of old American beer cans or tinny little radios in cheap plastic cases. The Deming Prize quickly garnered considerable prestige throughout Japan. The prime minister himself bestowed the honor in a solemn ceremony televised across the nation in prime time news reports.

Soon, Japanese buyers became even more demanding than U.S. consumers, and they wanted to buy from companies with a reputation for quality. The Deming program helped to spread the quality gospel across Japanese industry. Even more important, the stepped-up efforts of Japanese companies contending for the Deming honor over the past four decades helped to build an image of quality for Japan's products throughout the world.

If it worked for Japan, why not the United States? Starting with a few industrialists and members of Congress, with strong support from Commerce Secretary Malcolm "Mac" Baldrige, a push for such an American program began to take shape. Proposed legislation creating a program to honor U.S. firms achieving exceptionally high levels of quality failed to get through Congress in 1986 and was languishing in committee in 1987, when Secretary Baldrige, a dynamic businessman from Connecticut whose idea of weekend fun was to break a tough bronco, was killed in a rodeo accident. The tragedy gave impetus to getting Public Law 100-107 passed, and because of the late Secretary Baldrige's avid support, the proposed national program was named in his honor. The new legislation was tabbed "The Malcolm Baldrige National Quality Improvement Act of 1987."

～ HISTORY OF THE BALDRIGE AWARD: YEARNING TO REGAIN THE EDGE

The idea for what became the Malcolm Baldrige National Quality Award began with a Congressional delegation's search for ways in which the United States might regain its competitive edge in world markets. Don Fuqua, then a U.S. Representative (D-Florida), led the fact-finding pilgrimage that took the group to several Far East countries in January 1986.

In Japan, the representatives met with Kaoru Ishikawa, a well-known expert on quality and a counselor with the Union of Japanese Scientists and Engineers. Ishikawa told the delegation about the Deming Prize and its benefits to Japanese business, and that information stimulated interest among the delegation for creating a similar prize in the United States. The following June, an investigation into the pursuit of quality by U.S. businesses was begun by the House Subcommittee on Science, Research, and Technology.

The investigation included hearings that began in June 1986. Testifying witnesses included John J. Hudiburg, chairman and chief executive of Florida Power & Light Co., John Hansel, chairman of the American Society for Quality Control, and Myron Tribus, representing the National Society of Professional Engineers.

Hudiburg would prove to be a driving force in winning approval for the Baldrige Award. He and Sanford N. McDonnell, the chief executive of McDonnell Douglas Corporation at the time, led the effort to establish a foundation to provide funding for the award. Hudiburg recently retired as CEO to devote his time to promoting quality in U.S. businesses.

Legislation to create a quality award was introduced during the 1986 session of Congress but did not win passage that year. The legislation was reintroduced in January 1987, and additional hearings were held. Among those testifying at the second round of hearings was William Eggleston, vice president for corporate quality at IBM. After the hearings were concluded, the House Subcommittee on Science, Research, and Technology issued a report concluding, among other findings, that poor quality costs companies as much as 20% of their sales revenue, nationally.

On July 25, 1987, while legislation to create an award for quality in U.S. business was being debated, Secretary of Commerce Malcolm Baldrige was killed in a rodeo accident. Because Baldrige had been a strong supporter of a national quality award, his death proved to be a catalyst for passage of the legislation, and the award was named in his honor. Public Law 100–107 was signed by President Ronald Reagan on August 20, 1987, establishing the Malcolm Baldrige National Quality Award program for the United States.

～ *Baldrige Criteria Emphasize Customer Satisfaction*

Compared with Japan's Deming Prize, the Baldrige Award organizers broadened the criteria for the U.S. program, adding emphasis to such areas as customer satisfaction and service. Just studying the requirements to compete for the award is a lesson in how to achieve a corporate culture dedicated to total quality. To be contenders, for example, giant companies in major assembly industries, such as automobiles and computers, must bring their suppliers into their total quality efforts.

Even with award criteria booklets being distributed to many thousands of American firms, there was no guarantee that the necessary shifts in corporate culture would sweep American business and industry. How often has any government-run program achieved quick, effective results? The target here was not any particular industry or technology; it was aimed instead at achieving a major revolution across the entire business spectrum.

Organizers of the Baldrige Award management system carefully addressed this problem. To avoid the politics, rigidity, and lumbering pace of so many government programs, the Baldrige Award was set up as a public–private partnership. Management would be provided by the National Institute of Standards and Technology in Gaithersburg, Maryland, with assistance from the Malcolm Baldrige National Quality Award Consortium Inc. The consortium originally included the American Society for Quality Control in Milwaukee and the American Productivity and Quality Center in Houston (this latter group is no longer a member). The awards program also gets funding solely from American firms concerned with the global competitiveness of U.S. industry as a whole.

One of the most fortuitous decisions the organizers made in setting up the Baldrige Award program involved the selection of a director. As the United States faced its growing global competitiveness problem, Congressional leaders looked for a capable government organization that might take on the mission of supporting efforts to upgrade the quality and technology level of U.S. business and industry. The National Bureau of Standards (NBS) in Gaithersburg, just outside Washington D.C., seemed like an obvious choice because of its recognized technical competence and experience in supporting U.S. industry's needs for calibrating measuring devices as well as helping to develop standards. In the early days of computers, for example, NBS had successfully championed the ASCII code for data transmission. In line with the broadened mission envisioned by legislators, the name of the institution was changed to the National Institute of Standards and Technology (NIST). NIST also appeared to be an appropriate base for developing and propagating the criteria for the Baldrige program.

Curt W. Reimann, an analytical chemist who was deputy director of the National Measurements Laboratory and a former director of the NIST Center for Analytical Chemistry, was chosen to be director of the Malcolm Baldrige National Quality Award program. Reimann proved to be a man with a vision. He saw the Baldrige program as a base not just for transforming American business culture but for propagating new concepts for managing and operating organizations throughout society, including education, medical care, and the government itself.

In keeping with the customer focus of the total quality management approach, right from the beginning Reimann has sought input from business and industry on how to improve the award criteria and the process of selecting winners. Each year the program makes some modifications and improvements in the criteria as a result of these inputs.

The program designed careful procedures to ensure fairness in the selection of winners. It seeks new examiners each year to stimulate the flow of new insights and ideas in evaluating entries and providing feedback to all contenders to help them to strive toward total quality cultures.

All these efforts toward building a successful, respected national program are not an accident. Those who know Reimann say that it is not unusual for him to be at his desk as early as 5:30 a.m., and to work well into the evening, when he is not crisscrossing the country helping to preach the total quality gospel.

Despite the national quality program's growing success, each company considering entering the competition must decide whether it is worth the time and trouble. Companies such as Motorola, which won the coveted prize the first year of the competition in 1988, say it is not only worthwhile, it is essential for any company hoping to remain globally competitive. Others believe the process of becoming a strong contender is far more important than winning. Some companies admit that they entered the competition just to get the feedback from quality experts that is part of the Baldrige evaluation process. Other firms simply use the guidelines as a standard for evaluating their own efforts to boost total quality. There is little doubt, however, that the intense preparation necessary for a competitive entry can help achieve a genuine change in corporate culture, from the top to the bottom of an organization. The cultural shift certainly won't make life any easier for anyone involved, at least during the transition, but, if well done, the change to a total quality culture can dramatically improve the way a company is perceived by customers, suppliers, its own employees, and even competitors, no matter where they are. The results, say Baldrige contenders, soon begin to show up where it counts most in U.S. industry—the bottom line.

ᵔ Applying for the Baldrige Award Is a Rigorous Task

Becoming a Baldrige contender is no easy task. Just preparing to enter the competition requires a long-term plan that not only addresses the need to transform the organization's culture based on total quality principles, but also includes methods for tracking the success (or failure) of continuous improvement efforts. All these results do not have to be included, but if the entry is considered good enough to be a contender for the top prize, a team of examiners will visit selected plants or offices to verify claims and to clarify any vague or confusing points. These quality experts will insist on seeing actual evidence of steady improvement over a period of years.

Awards are given in three categories: manufacturing companies or their subsidiaries, service companies or their subsidiaries, and small companies (with no more than 500 employees) engaged in either manufacturing or services. Up to two awards may be given in each of the three categories. Only companies located in the United States are eligible, and they may be private or public companies.

The criteria on which the judging is based falls into seven categories:

1. the ability of a company's leadership to establish a culture that emphasizes quality as a goal
2. a company's efforts to collect and analyze information to improve quality
3. a company's effectiveness in incorporating quality into its business plans
4. a company's utilization of human resources to achieve quality
5. the effectiveness of a company's quality assurance control programs
6. quantitative measures of the results of those programs
7. customer satisfaction, on which there is heavy emphasis

The award's criteria are gaining a reputation in the business world as the right way to define quality. They range far beyond the narrow view of quality held by most American managers just a few years ago—such simple measures as acceptable quality levels or mean time between failures. These are important, of course, and the award requirements include evaluations of methods used for quality assurance of products and services along with the quality levels that result.

The Baldrige program has encouraged organizations and even regions and localities to make use of the evaluation criteria to set up a sort of mini-Baldrige award programs of their own.

The program has revised its scoring methods over the course of the national award. In 1990, each category was assigned a number of points, with the total equaling 1,000. A more elaborate scoring scheme was devised for 1991 and 1992. Examiners emphasize quality results, not just systems and programs put in place to achieve quality. The scoring is done by a Board of Examiners made up of more than 200 quality experts from industry, universities, and professional and trade organizations.

∿ On-Site Visits Lead to the Final Choices

At least four examiners review and score each application, which can amount to 75 pages for large firms (plus appropriate supplemental sections in some cases) or 50 pages for small companies. A team of five examiners selects high-scoring applicants for on-site visits. The team chooses sample sites for companies with multiple sites. The judges' panel of the Board of Examiners recommends which companies should receive the award, with the Secretary of Commerce making the final selection of the award recipients.

Originally, the identity of those chosen as winners was not publicly announced until the day of the ceremony in late fall, but starting in 1990 this policy was changed so that winners could be informed a few weeks before the ceremony. In 1988, President Reagan personally presided over the ceremony and gave out the awards, and in each year since then President Bush has made the presentations.

The Baldrige Award program encourages winners to publicize their receipt of the prize, thus giving visibility to the award. Although applicants are not required to disclose proprietary information about their operations, winners are expected to share information about their quality strategies with other U.S. organizations. One factor considered in choosing winners is that they can serve as worthy models for emulation by other American businesses.

All applicants, whether final contenders on not, receive a written summary of their strengths and areas for improvement the examiners find in each company's quality programs. Applications and guidelines are generally available in December, with entries due in early April. An entry fee ($50 in 1991) is required for each application for initial verification of the eligiblility of the applicant. Application review and the site visit selection process has run from the April deadline through the end of September. The awards ceremony is scheduled for October or November, and feedback reports are distributed by November or December.

Application fees in 1992 were $4,000 for large manufacturer or service companies and $1,200 for small companies. Proportionately higher fees may be required if additional information is needed to describe the quality process for distinctly different product lines, service lines, or business units (in 1992 a fee of $1,500 was set for each

~ THE BALDRIGE AWARD AND THE DEMING PRIZE: HOW THEY DIFFER

Although there are many similarities between Japan's prestigious Deming Prize and the Malcolm Baldrige National Quality Award in the United States, there are also some important differences, according to Curt W. Reimann, director of the Baldrige program at the National Institute of Standards and Technology. The Deming Prize, begun in 1951, is named for the legendary American quality consultant W. Edwards Deming, who helped the Japanese overcome an international reputation for shoddy merchandise. The Baldrige Award, which was first issued in the United States in 1988, is named for the late U.S. Secretary of Commerce Malcom Baldrige.

Both are annual award programs that are intended to raise the quality levels of industry in the respective nations. Some of the mechanics, such as the submission of applications and site inspections by examiners, are similar. The categories differ somewhat, however. The Deming, like the Baldrige, can go to large or small manufacturing enterprises or to a division of a large firm. There are also Deming Prizes for individuals and quality control awards for individual factories.

By contrast, the Baldrige Award has a separate category for service organizations and may even include one for government departments in the future. Reimann says this is in line with the overall objective of the U.S. program of disseminating as widely as possible in American society a value system based on total quality management concepts. The areas of customer satisfaction and services are also weighed more heavily in the Baldrige Award program.

The Baldrige evaluation is more results-oriented than prescriptive. The Deming criteria are more specific about quality control and problem-solving methods. Companies planning to contend for the Deming Prize must contract with quality specialists associated with the Union of Japanese Scientist and Engineers (JUSE) in order to raise their quality standards to levels considered worthy of the award.

By contrast, Baldrige examiners have no defined consulting relationship with the potential applicants. Baldrige examiners are carefully selected from a pool of applicants each year, thus encouraging some turnover and fostering wider participation.

The Baldrige process involves both a competitive and a qualifying system, whereas the Deming Prize is a qualifying system based primarily on the prize criteria as judged by the JUSE counselors. In the U.S. system, experts from industry, government, and universities match applicants against the best in the world, and only those entrants judged to be appropriate national models are chosen for final qualification.

Unlike the Deming Prize approach, the Baldrige Award procedures require a built-in process of continuous improvement through an annual cycle of updates. Also, an ever-widening range of organizations is being encouraged to cooperate to reach a consensus on any changes in the Baldrige criteria. Building the program around a seven-category framework ensures some basic continuity, however, and only fine-tuning is done in Baldrige requirements from one year to the next.

---

supplemental section). Fees have varied since the program started, so it is important to check them before entering. The applicant pays expenses associated with site visits.

It is certainly possible to achieve a transformation to a total quality culture without competing for the Baldrige Award. The process of studying the Baldrige criteria, however, and applying them across the organization to measure achievements and shortfalls, can serve as a powerful force for change. This book offers plenty of ideas for turning a company into a far tougher, total quality competitor, whether or not it enters the Baldrige competition.

Chapter 2

The Long, Hard Road to the Baldrige Award

Competing for the Malcolm Baldrige National Quality Award seems to be no one's idea of a good time. At the very least, the process consumes huge amounts of time and, in some cases, money. Worse yet, even the most hardened U.S. executives have been humbled by confronting the scores of defects, large and small, that make total quality so elusive for their companies.

So why is it that the Baldrige Award has apparently captured the imaginations of so many executives? One reason is that top managers at more and more U.S. companies realize that in order for their companies to thrive—or even survive—in the new global economy, they must become truly world class. Increasingly, they see the Baldrige guidelines and criteria as powerful tools for reaching that status.

Any U.S. company, no matter what size or in what industry, whether in manufacturing or services, can compete. Up to six awards can be presented each year, two each in the categories of manufacturing, services, and small companies (fewer than 500 employees). Companies that have entered have learned that the process fully lives up to its reputation as an exhausting exercise. Where entrants once hesitated to reveal their candidacies (perhaps fearing they would be judged critically if they failed to win), the Baldrige competition now is "perceived as more like competing in the Olympics," according to Curt W. Reimann, director of the Baldrige Award program at the National Institute of Standards and Technology (NIST).

Beginning in 1990, entrants could pay a fee (deductible from the entry fee) to learn whether they met eligibility rules before submitting a full-scale entry, which could run 75 pages or more for a large company. In rare cases, an entry will be turned down for procedural reasons. No Motorola division could enter in 1990, for example, because the company as a whole had won the honor within the preceding 5 years. If a division won and its sales were less than 50% of the corporate total, however, the 5-year restriction would not apply to other divisions, so even if one division won, other divisions in the same company could enter in subsequent years. Since such details of the rules have varied from year to year, it is critical to check the latest rules for entry in the current application guidelines.

One way a company can determine its chances of winning before actually entering the competition is to conduct an internal quality assessment using the current guidelines, which spell out the rules, evaluation criteria, and method of scoring. A new set of guidelines for the competition is available from NIST early each year. Some adjustments are made in guidelines each year, based on feedback from entrants and potential candidates.

Some larger companies rate their chances of success by pitting the overall quality of various internal divisions against each other, then naming the winner as that year's Baldrige entry. Since many categories require evidence of improvement over time, even companies that decide to enter will sometimes work at improving their potential score for years before deciding to "go for the gold."

How high a score does a winner need? According to 1990 rules, a competitor could receive a maximum of 1,000 points, divided among seven major categories. According to Dr. Reimann, however, winners of the early competitions scored in the high 600s to low 700s. That range could rise, he points out, if stronger entries are received as the program progresses.

The judging itself appears to be fair but tough. Entries are scored by a selected group of more than 200 Baldrige examiners—quality experts drawn primarily from industrial and service organizations but also from government agencies, the military, and academia. During the company evaluation, examiners report that they are especially watchful for three major indicators of a successful quality program:

- *Approach.* What strategy and methodology does the organization use in attempting to achieve world-class quality?
- *Deployment.* What resources are being applied, and how widespread is the quality effort throughout the organization?
- *Results.* Is there convincing evidence of sustained improvement?

Entrants tend to score within a fairly narrow range on approach, and a few companies trip up in deployment (usually because of too much focus on one area, such as a particular product line or just manufacturing operations). It's the last factor—results—that most quickly separates the noncontenders from the potential winners, say the examiners, since the criteria are heavily weighted toward customer satisfaction. Moreover, scoring high on results requires convincing data that show steady improvement over time, internally and externally. Even with a good internal program, it takes time for the results to show up. If such evidence is missing or sketchy, the overall score will suffer.

One thing is clear: No company can win simply by submitting a slick entry. Finalists in past competitions were subjected to 350 to 400 hours of evaluation by several quality experts, including site visits. Quantitative results weigh heavily in the review and judging process, so competitors must prove that their quality efforts have resulted in sustained improvements.

Several examiners completely review every application in order to prepare comments on how total quality might be improved. Thus, even applicants that are not selected as finalists get valuable feedback.

Examiners develop individual percentage scores for each category. For example, an 80% rating for the category of customer satisfaction (which is worth a maxi-

---

∾ How to Get Copies of the Baldrige
   Award Guidelines

For information about the Malcolm Baldrige
National Quality Award competition, or for a free copy
of the guidelines, write or call:

Malcolm Baldrige National Quality Award Office
United States Department of Commerce
National Institute of Standards and Technology
Gaithersburg, Maryland 20899
Telephone: (301) 975-2036
Fax: (301) 948-3716

---

mum of 300 points) would result in a score of 240 points. Any differences in scores within a given range are negotiated by the group of examiners to reach a consensus. To achieve a 100% rating in one item, there must be concrete evidence that excellence in that category has been sustained over time.

In rating entries, scores in any one category tend to be linear up to 50%; beyond this, the scoring becomes more logarithmic, suggests one senior examiner, meaning that the higher the score, the harder it becomes to justify an increased score. A score from 50% to 90% indicates a very strong quality program with a sound plan, wide deployment, and good results. To reach the 90% to 100% range, however, management's commitment to quality must be demonstrated by a sustained high level of excellence through multiple business cycles. Groups of up to six examiners visit the sites of finalists to verify claims in high-scoring entries.

Rather than being prescriptive, notes Reimann, the criteria are designed to allow each organization to meet the Baldrige requirements by making adaptations to its own individual corporate culture. While the overall approach has remained the same from year to year, there have been some shifts and some fine tuning in scoring emphasis.

If the Baldrige criteria are most focused on one particular area, however, it is the way in which companies interact with their customers. That has been one of American industry's weak spots, feels Richard Hoff, a senior quality consultant from OmniTech International Ltd. in Midland, Michigan. The lack of good measures of customer satisfaction were clear in entries for the first years of the competition, but he claims to see evidence of improvement in this area with each succeeding year.

Given the huge commitments of time and energy, many companies still ask themselves: Why should we enter? Sources close to the Baldrige process offer several compelling reasons. One of the most important is the avid support for the Baldrige Award among such giants as General Motors, Motorola, Xerox, Westinghouse, IBM, and Texas Instruments. Small and medium-sized companies are coming to recognize that a Baldrige candidacy could give them a powerful edge in competing for business from these big companies.

Motorola has even notified thousands of its suppliers that in order to keep doing business with Motorola, they must come up with plans and a target date for entering the competition. Why the ultimatum? Motorola claims that it cannot reach the stringent quality levels it has set for itself unless all its vendors strive for similar standards.

Many companies feel the need to become world-class players in a shrinking global economy, but before a company can reach that status, "it must understand what world-class really means," says senior Baldrige examiner Arnold Weimerskirch, director of quality for Honeywell in Minneapolis. Benchmarking against highly competent organizations is an emerging science in the United States, he says, but many companies need to put more resources into gaining enough knowledge to do this effectively. As the Baldrige examination process becomes ever more precise and refined, it provides a clearer and more uniform definition of what world-class quality really means.

The Baldrige program also is a powerful training and education tool, according to Harry Kenworthy, who heads quality at Rogers Corporation in South Windham, Connecticut. Because every Baldrige entry receives feedback from experts, adds Reimann, the cost of building toward world-class caliber is minuscule—especially when compared with a program like the Deming Award in Japan, which may require a company to spend several years and hundreds of thousands of dollars to bolster its quality systems.

In the past, quality programs were often relegated to an obscure unit which thrived on paperwork, buried somewhere in the organization. The Baldrige competition has radically changed this, claims Reimann. The contest forces a much broader quality effort that requires each contender to measure its own efforts toward an enterprise wide continuous improvement process against the best in the business.

The fixing of a corporate mind-set, in fact, is probably the most important point of all. Seminars and conferences certainly have their place, notes one Baldrige examiner, by explaining such topics as just-in-time manufacturing, statistical process control, shorter set-up times, lower inventories, and employee empowerment. But unlike the Baldrige competition, which embraces and combines these and many other elements of quality under a single broad umbrella, such approaches carried out in isolation are of questionable value in achieving the cultural shift that is demanded by an authentic total quality environment.

No longer is quality seen as just an added cost of making acceptable products. The new thinking on total quality affects far more than products and processes. It aims to improve relationships as well—with customers, suppliers, employees. It promises to transform organizations. The cultural change can only be accomplished with a lot of hard work. It can't be added on to existing programs or bought from a consultant. But the companies leading this quiet revolution are discovering that the new pursuit of quality cuts costs, too. Ultimately, the new thinking could change American society.

To join the total quality revolution, and to become a legitimate contender for the Baldrige Award, requires commitment from every level of the organization, starting at the top. The effort must engage all employees, each of them continuously striving to meet ever-higher standards of excellence. The heavy emphasis on satisfying the customers of the organization's products or services ensures that judging will be based on substance rather than form. The required evidence of steady improvement over time will rule out those organizations treating the total quality approach as another fad offering a magic pill to solve deep-set problems that instead require a cultural transformation over an extended period.

The effort to meet Baldrige Award requirements must not be seen by employees as another "campaign of the month" or even "campaign of the year" imposed by management. "The company must demonstrate it has at least a 3-year horizon," says Reimann, "with a long-term framework for the orderly and consistent discussion of total quality management throughout the organization."

∼ A Catalyst for Revolutionary Change

The goals of the Baldrige program, Reimann adds, go far beyond the public relations value of a national quality award. In fact, this program is seen as a catalyst, he says, to bring revolutionary change—not just to the U.S. business community but to American society. The award is aimed at boosting all levels, so that Americans can take pride in the quality of U.S. goods and services and so that the global market begins to see the United States as the source of the world's finest goods and services.

To reach such lofty objectives, wide participation is essential and the program must be viewed as a fair and objective appraisal. One way in which the program gained credence was by enlisting a wide range of professional organizations as well as individual companies to organize it. Scores of recognized quality leaders and organizations helped formulate the criteria, evaluation procedures, and training curricula for the award, including the American Society for Quality Control in Milwaukee, which now helps the government manage the program, and the American Productivity and Quality Center in Houston.

Drafts were circulated to numerous quality specialists and input was sought until consensus was reached. From some 150 nominations in the first year, about 100 examiners from a wide range of organizations were chosen to serve on a Board of Examiners. That number has steadily expanded. The program invites new nominations for examiners each year and repeats the selection procedures, thus broadening the participation of companies and organizations in the program. In selecting examiners, Reimann explains, the Baldrige program particularly seeks those involved in their regions, communities, and professional and trade groups.

The hope is that the concepts behind total quality management will expand into other areas of American society, including schools, healthcare facilities, and other activities outside the commercial sphere. A section in the law setting up the Baldrige Award would allow the Secretary of Commerce to enlarge categories so that government departments might be included in a special category. Reimann says that this is being actively considered for the future.

Even though Congress passed a law establishing the program, a tribute to its acceptance is that funding comes from the private sector. The $10.5 million needed to get the award program rolling was donated by private industry in less than a year, says Reimann. Continuing activities are being supported by entry fees.

Although the philosophy behind the Baldrige Award will not change, there will be some fine tuning each year, based on industry feedback and a desire to focus on areas where the need seems greatest. Although there has been some effort to simplify the process of filling out entry forms, Reimann points out that one person over a weekend filled out the entry of Globe Metallurgical, the first winner in the small company category in 1988. The criteria will be sharpened each year as well, "to help differentiate the excellent from the merely very good," explains Reimann.

For management, key criteria include management style and how well it is attuned to continuous improvement. "Management should act as a servant to the organization, rather than a ruler; a coach, not a dictator," he says. Converting an organization to a new culture takes time and leadership.

"Empowering is a term we use commonly," says Reimann. Moving more employee authority and responsibility out to the customer interface changes the way people in the firm interrelate. At the same time, if front-line employees are to be given more authority, it can't be done in such a way as to reflect badly on middle management, Reimann says. A group effort should be emphasized, with middle management getting recognition for the efforts of group members to improve operations or to better serve customers.

"If every business unit is challenged to make significant improvements each year, the firm will need ideas from all over, including from employees. Management can't just walk in every year and tell them how they can make big performance gains.

"Global competition is a moving target," states Reimann. But too many U.S. companies have been attuned to damage control, he believes, essentially responding to problems rather than preventing them in the first place. Judges look at such factors as closeness to customers and planning based on analyses of how products and services are used in the marketplace.

∾ The Baldrige Examiners: Tips from the Top

The Baldrige competition examiners—the more than 200 men and women who are charged with defining quality for U.S. industry—are a diverse group of quality specialists who measure every application against the Baldrige criteria to provide the judges with enough information to determine which companies are good enough to rate the all-important site visits. The examiners also use the initial analyses to provide feedback on strengths and weaknesses that a company can use to upgrade its quality efforts.

Being a Baldrige examiner is no easy job. Applicants must agree to volunteer at least 10 working days a year on the program just to be considered. In 1990, examiners, senior examiners, and judges were selected from almost 600 applicants. Examiners serve for one award cycle only but may reapply for assignment the following year. The selection process considers background in the quality field but also is aimed at striking an industrial, as well as a geographic, balance.

Each examiner undergoes 3 days of intensive training to learn a uniform approach to judging and scoring entrants. Examiners work together as teams, with a senior examiner responsible for building consensus on final scores. There are only nine judges; their job is to review scores and scoring profiles in order to choose applicants for site visits. After reviewing site-visit reports, the judges recommend potential award recipients to the NIST. Award-winners, which are expected to serve as ambas-

∾ Six Steps to the Baldrige Award

1. Internal assessment

   Although it is not required, many companies find it worthwhile to conduct an internal review of the company's quality standards and achievements, usually using the actual Baldrige criteria as guidelines.

2. Determination of eligibility

   All applicants must submit an Eligibility Determination Form (available from NIST), along with a nonrefundable $50 fee; deadline for submission will be announced by NIST. The form is intended to identify ineligible applicants.

3. Application submission

   Once NIST confirms a company's eligibility, applicants submit the full completed application, together with the appropriate entrance fee. The application guidelines provide detailed instructions.

4. Review by examiners

   All entrants get a first-stage review by four or more members of the board of examiners, led by a senior examiner. Candidates that pass proceed to an additional review that determines which candidates receive a site visit.

5. Site visits

   The few candidates still in the running at this point are scheduled for a personal inspection of the site by at least five members of the board of examiners and a senior examiner. A panel of judges makes award recommendations and the Secretary of Commerce makes the final decisions.

sadors for quality in the United States, must also be approved by the Secretary of Commerce.

Rigorous conflict-of-interest guidelines prevent any connection between an examiner and an applicant, whether by employment, client status, significant stock ownership, or competitive relationships, according to Reimann. It also is a violation of the rules for an examiner to even inquire about the status of an application other than those assigned to that individual.

Not surprisingly, many of the examiners have strong opinions on the subject of quality and on how U.S. companies often fail to measure up. "American managers are capital-happy," says one examiner. "They're geared more toward buying tools and techniques."

What companies really need is to do a better job of using the resources they already have, suggests former examiner Jack ReVelle, a quality specialist with Hughes Aircraft Company in Los Angeles. Learning more about such methods as statistical process control and experimental design, he maintains, could greatly improve the effectiveness of existing equipment in virtually any American plant.

Other examiners complain that many executives fail to understand that the lack of quality is rarely attributable to production workers. "It is a management problem, and it requires a cultural change in U.S. companies, led by top management,"

∽ WHAT IT WILL TAKE TO WIN THE BALDRIGE AWARD

Following is a summary of the judging criteria for the Malcolm Baldrige National Quality Award:

Leadership. The senior management's success in creating and sustaining a quality culture.

Information and analysis. The effectiveness of the company's collection and analysis of information for quality improvement and planning.

Planning. The effectiveness of integration of quality requirements into the company's business plans.

Human resources utilization. The success of the company's efforts to utilize the full potential of the work force for quality.

Quality assurance. The effectiveness of the company's systems for assuring quality control of all operations.

Results. The company's results in quality achievement and quality improvement, demonstrated through quantitative measures.

Customer satisfaction. The effectiveness of the company's systems to determine customer requirements and demonstrated success in meeting them.

---

contends former examiner Gerry Lenk, a Monroe, Connecticut, consultant who once headed quality programs at Pitney-Bowes in Stamford. "And in this country senior management has the job-life expectancy of a nose tackle in the NFL." For too many companies, the real focus is on quarterly earnings, management bonus systems, and pressures from institutional investors looking for quick paybacks.

Lenk and many of his colleagues laud the efforts of Motorola and other Baldrige winners to share information about their total-quality efforts with the rest of U.S. industry. Participative management programs, which provide for bonuses—up to 40% of annual salary—for meeting tough targets, also provide powerful incentives for continuous improvement in some companies.

In any event, says Lenk, management must be willing to make some small initial investments while realizing that it may take a couple of years to see returns. When efforts do begin to pay off, he suggests that perhaps 40% of the gains be plowed back into the continuous-improvement process.

There is no better way to gain a sense of direction in moving toward total quality than to study the scoring applied to the seven major categories of the Baldrige applications criteria. In 1990, a perfect rating of 100% in every category would result in a maximum total score of 1,000 points. Since actual winners have only had scores in the high 600s to the low 700s, even the best U.S. companies still have a long way to go to reach the top rank in world-class total quality.

The categories and their point scores in 1990 were:

- leadership (100 points)
- information and analysis (60 points)
- strategic quality analysis (90 points)
- human resource utilization (150 points)
- quality assurance of products and services (150 points)

- quality results (150 points)
- customer satisfaction (300 points)

A more intricate scoring system, assigning percentages for each item based on approach, deployment, and results, was instituted in 1991. It is important to check the current applications guidelines for a full description of the scoring method that will applied in subsequent years.

∼ To Impress the Judges: Save the Razzle-Dazzle

Okay, so you've decided it's worth the effort to go all out to try to win a Malcolm Baldrige National Quality Award. Before you start ordering the bubbly, however, there are a few things you should know about the process of picking winners.

Becoming a contender depends on what the judges see in the entry forms, but don't assume that fancy writing and multicolor charts are the road to victory. "Examiners want simple, precise documents," says E. James Tew, a senior examiner for the Baldrige program. Tew is the manager of quality assurance operations for Texas Instruments Inc.'s Defense Systems and Electronics Group in Dallas. "You can leave out the razzle-dazzle," Tew advises.

At NIST, Program Director Reimann echoes this recommendation, indicating that beautiful, slick presentations made up by professional promotional organizations are less likely to catch the judges' attention than simple entries in the words of those carrying out an organization's quality programs. Tew explains that the judges look for a simple, convincing account of a well-conceived, total quality program. What kind of a submission catches their attention? Major scoring criteria, as detailed by Tew, are:

- *Soundness.* Does the overall approach to quality make sense?
- *Deployment.* Are there sufficient resources and people to carry out the program?
- *Results.* Are the results anecdotal, or is there a valid, long-term program with measurable improvements over a period of time?

Once they have chosen the contenders, the examiners may spend days flying from plant to plant to determine whether the claims made in the entry are valid. One sign of the strong support U.S. industry has for the Baldrige award is that the examiners, such as Tew, with the support of their employers, work some 20 days or more a year without compensation for their efforts. Although this means an occasional day away from their regular jobs, to tour factories during the week for example, it also calls for many evenings, lunch hours, even weekends or holidays, according to Tew.

Scoring is by consensus, Tew explains. Each of the team of quality specialists develops scores individually. "It's surprising how close these come out," according to Tew. But if there are any discrepancies, the examiners, coordinated by a senior examiner, arrive at a single consensus score.

The award ceremony takes place in late fall (usually November). If your company makes it through this rugged obstacle course, a video showing something about how you achieve superior quality throughout your operations will be one of the highlights of the ceremony. Then comes the hardware: elegant crystal statues encasing medallions, each inscribed with the name of a winning company. The President of the United States personally congratulates the winners.

At last, it's time to celebrate. You and your hard-working team can go back to the factory floor and pop those magnums of champagne. Each year, a few more fine American companies make it through to win a coveted Baldrige Award. You'll find the stories of all 12 winners through 1991, as well as descriptions of the superb efforts of scores of other firms to become winners, in Part II of this book.

Chapter 3 ~

Industry's Oscar: The Payoff Can Be Worth the Effort

In spite of the difficulty of preparing for and entering the Malcolm Baldrige National Quality Award competition, requests for copies of the applications guidelines have been steadily rising, topping 180,000 in 1990 (information on where to obtain copies is presented with the list of Winners of the Baldrige Award and on page 15). Curt Reimann, director of the program at the National Institute of Standards and Technology (NIST) believes the huge demand for entry booklets demonstrates that plenty of companies hope one day to win industry's version of Hollywood's Oscar and to get the personal congratulations of the president of the United States. But there are other reasons for the demand as well.

A company doesn't have to enter the competition in order to get copies of the Baldrige guidelines. The stringent criteria can be useful just to evaluate efforts toward improved quality and competitiveness. Some companies go further, using the guidelines to set up internal award programs for divisions or suppliers. Some states and regions have set up competitions built around the national Baldrige guidelines. In some cases, pressure from big customers is a factor. Corporate giants, most with vigorous programs of their own to go for the award, may press their suppliers to develop plans for entering the competition. Or they may insist that vendors at least adopt total quality practices, and the Baldrige criteria supply the language and focus for such efforts.

Thus, the benefits of the Baldrige program go far beyond increasing the global competitiveness of the winners. A process has been set in motion to ensure top-quality performance by thousands of American organizations, whether or not they enter.

If actually winning the competition is so difficult, why bother with the intense effort required? The record already shows that winning a Baldrige Award can prove sweet indeed. Want fame? How about profiles in *Fortune,* appearances on the "Mac-Neil-Lehrer News Hour" and "The CBS Evening News" and coverage in *USA Today, The New York Times,* and other major newspapers and magazines all over the United States? This is the publicity that Motorola received after winning a Malcolm Baldrige National Quality Award in 1988.

Motorola's executives continue to field a deluge of requests to address business gatherings. For many months after the company won the award, as many as 50 companies flocked to sessions at a Motorola auditorium at its Schaumburg, Illinois, headquarters to hear officials espouse the company recipe for quality. Milliken & Company, Spartanburg, South Carolina, charges $300 a head and still books some 70 executives and managers to attend day-long quality seminars a couple of times a month. Federal Express gets large numbers of attendees at sessions that start with a late-night look at its mammoth sorting center in Memphis, where packages from more than 100 aircraft are transferred for overnight delivery. The company follows up the next day with presentations on its excellent total quality systems.

Or how about fortune? Globe Metallurgical, Inc. of Beverly, Ohio, a $120-million maker of metal alloys, saw its business grow 10% after winning one of the three inaugural Baldrige honors, when it had been expecting to sustain a flat year. Globe got new quote requests from as far away as Australia, says Kenneth E. Leach, a former Globe vice president, and its existing customers began to give the company more business.

Indeed, fame and fortune do await those that win the Baldrige Award. Their celebrity reaches out to international markets as well. The quality-conscious Japanese show particular interest in Baldrige winners. The award places U.S. winners in the class of Japan's best in terms of quality, and the high status of the Deming Prizes there adds to the prestige.

In the United States, the biggest guns in industry—such as IBM, Ford, General Motors, Hewlett-Packard, Harris Corporation, and AT&T—view Baldrige winners as very special, indeed. "We think those are certainly the folks we ought to be doing business with," says Gary Bobleter, director of materiel at Harris Semiconductor, referring to Baldrige winners.

Winning the Baldrige Award does far more than give companies a logotype to print on their business cards and a line to use in sales brochures. It catapults each winner to the lofty status of "quality guru." Baldrige winners are not merely encouraged but are expected to spread the gospel of quality throughout industrial America. This includes speaking at conferences, hosting tours of their facilities, and promoting the award in advertising and public relations.

The winners don't seem to mind bearing such a burden. As it turns out, "Companies have been going far beyond what what was asked of them," says Reimann. "I think there's a pride factor that's enormous, a patriotic factor that's enormous, and a marketing factor that's enormous."

It is important to note, however, that companies do not compete for the award for the marketing benefit alone. The reason companies seek the award, says Reimann, "is really rooted in a deep belief in quality." Affirms Wayland Hicks, executive vice president with Xerox Corporation, whose Business Products and Systems Division was a winner in 1989: "Our whole pursuit of the Baldrige Award was to make ourselves a better total quality company."

A Baldrige contestant must go through a rigorous self-examination of its quality systems. Contestants must look at virtually all facets of how they do business and correct any deficiencies they may have—from minimizing the time it takes to answer customer inquiries to diminishing outgoing defect rates. "We think the day we apply for that thing we will have already recognized significant share increases in our business," says Michael Hartnagel, director of information systems at DuPont Electronics.

Perception, however, is everything in the marketplace, and aspirants say winning the Baldrige Award could close any gap that may exist between the perception of quality and its demonstrated reality. "There is no question that it would have enormous market potential for us," says Michael Werner, manufacturing vice president at GCA, a division of General Signal. GCA makes integrated circuit production equipment and planned to enter the award competition in 1992. "In the competitive industry that we're in, being singled out as someone who meets these standards has a marketing value that is unparalleled."

American integrated circuit (IC) lithography equipment makers, like GCA, are among those that could benefit greatly from a Baldrige award. While the general market consensus is they lag behind Japanese firms, such as Nikon, in quality, this is not true, they say, and winning the award could settle the problem. "It would be great," says Raymond Campbell, president of Ultratech Stepper, a unit of General Signal. "A Baldrige winner has the bloody Good Housekeeping Seal of Approval stamped on its forehead!"

∾ Different Ways to Benefit

The Baldrige award will not benefit all companies and segments of industry equally, as far as marketing is concerned. Marketing consultants, such as George Cohan, president of Q/Mark in Chicago, a unit of the Starmark Co. advertising agency, say it provides less value to a market leader that is already widely known for its high quality and more value for a firm that has a weaker market position.

Businesses that serve regional or nonindustrial markets also may find the award lacking in market value because the Baldrige reputation is not as widely known outside the circle of major global corporations and their suppliers. For Globe Metallurgical, however, it has provided enormous benefit, which the company can measure on the bottom line. Prospective customers that had never heard of Globe before it won the Baldrige Award now are steady buyers. "It's lifted us from about total obscurity to a position in which we enjoy national and international recognition," says Globe's former vice president Leach.

As for Motorola, the largest and most widely known of the early winners, the impact is probably not as dramatic. Motorola officials could not quantify how much additional business the Baldrige Award brought in, and they emphasize that the purpose of competing was not to win an award but rather to help focus continuous improvement efforts within the company. They feel that the total quality approach is essential for any company that hopes to remain globally competitive. Still, some Motorola executives do suggest that benefits will occur over the long term. "We've asked some consultants and they say that we're planting a mind-set," says Scott Shumway, a Motorola vice president and director of quality for the semiconductor group.

Ever since the company won the award, all Motorola's top executives—its chairman, vice chairman, president, sector and group general managers, and other senior managers—have been helping to spread the quality gospel across the United

States. In 1989 alone, these corporate leaders lectured on quality to close to 400 gatherings, from business association meetings to quality congresses.

The Baldrige briefings, Motorola's monthly day-long quality seminars, each pack as many as 150 executives from 50 companies. For those companies that want personalized guidance on quality, Motorola hosts individual all-day quality powwows with its top executives and managers. Within the first year of the program, Motorola provided these one-on-one sessions to well over 100 companies.

In these presentations, Motorola reveals how it achieves near perfection in almost everything it does—in other words, how it won the Baldrige Award. Beyond fulfilling a patriotic duty to its nation, these sessions pay priceless dividends. They bolster relationships with existing and potential customers. Motorola's Shumway reports that the Baldrige Award has bought the semiconductor group especially tight relationships with at least six major companies: IBM, Rockwell International, GM's Delco Electronics, Tektronix, and Mentor Graphics. At one company, which has virtually copied Motorola's quality system, "almost any level of Motorola can get on the phone and call at any of the divisions, and they'd be welcome," says Shumway. "Our managers and salespeople are talking to individuals that they'd never been able to talk to before."

The relationships that come from being a Baldrige-certified example of quality extend to the level of company president, adds John Ristow, director for corporate quality assurance at Tektronix, Inc., and a Motorola disciple—making its benefit particularly significant. "From a marketing standpoint," says Ristow, "what better advantage can you have than to be talking to the president of the company?"

In addition, Motorola spreads its Baldrige message far and wide. Call in its own people to hear about one of its advanced microprocessors, and there will also be an earful on the Baldrige. Visit Semiconductor Product Sector's headquarters in Phoenix, and one will see on a pedestal in the center of the lobby a full-sized replica of the gold Baldrige medallion. Motorola employees place a gold foil sticker with the Baldrige logo on each letter that goes out, and a photograph of the award is a feature of its annual report. Even shipping cartons carry the Baldrige logo.

Motorola thanked its employees, customers, and suppliers in full-page ads in the *Wall Street Journal* and several dozen local newspapers from Miami to San Jose. It staged quality day bashes at its manufacturing facilities around the world, with local dignitaries reenacting the Baldrige ceremony with coverage by the local press. It kicked off a multimillion dollar image-building advertising campaign using its continuing pursuit of quality—and the Baldrige prize—as the theme.

Other companies have heavily promoted their victories as well, particularly Xerox, GM's Cadillac Division, and Federal Express. At the same time, they tie the Baldrige program to a revival of quality in American industry. The enthusiastic support by such giant corporations for the national push toward total quality has reverberated powerfully across the business community. Motorola's push for all its suppliers to come up with plans to compete for the Baldrige Award, and more subtle suggestions from such companies as IBM and Westinghouse, have had a telling effect.

The reaction is summed up succinctly by the head of a small electronics components company. For capacitor maker AVX Corporation, preparing to enter the Baldrige competition has become a requirement of doing business. "You've got a major customer or a major potential customer who says that you've got to do it. That's the end of the discussion," states Marshall Butler, AVX chairman.

Undoubtedly, the award's marketing value will be redefined over the next several years. But given the momentum so far, this appears to be slated to become an award that virtually every company covets. Still, the national efforts toward total quality have a long way to go in the United States before they match the prestige of Japan's program. When the Baldrige Award presentations are shown on prime-time television and are trumpeted in front-page headlines in American newspapers, as are the Deming Prizes each year in Japan, then the hard work of many hundreds of American executives, workers, and quality specialists will have paid off.

Part II ∿
Total Quality in Business and Industry

Chapter 4 ⌒

Motorola

One of the First Baldrige Winners

Motorola, Inc. has an unusual problem. In 1988, the inaugural year of the Malcolm Baldrige National Quality Award, Motorola was one of the first large companies to win the coveted prize. So what's next? While many other U.S. companies scurry to put together credible quality programs worthy of Baldrige contention, Motorola has to wait 5 years just to be eligible to compete again. That means that the quest for the award can't serve as an immediate driving force toward quality, as it does for corporations that have never won the award.

A Baldrige winner, of course, must agree to share information on its approach to quality with other American firms. Motorola has certainly been doing that, along with other past winners, by sponsoring seminars and describing its methods to scores of suppliers and even competitors and, of course, trumpeting its victory to its customers and prospects. But glowing in the spotlight of success could be a prescription for complacency. Not so at Schaumburg, Illinois–based Motorola, says George Fisher, president and CEO. The Baldrige competition, in fact, was only considered a way station on a route to an even tougher challenge. "Yes, we won the Baldrige Award," says Fisher, "But all that means is that we are on our way with an effective process aimed at achieving total customer satisfaction. We still have a long way to go."

International competition, particularly from Japan, is getting steadily tougher. So, Fisher explains, the company is struggling to reach the seemingly impossible goal of six-sigma quality in everything it does by 1992. In short, six sigma means that Motorola wants its quality program to hit its targets 99.9997% of the time: that is, no more than 3.4 defects per million units. By introducing new products and carving out new niches, Motorola has doubled in size roughly every 5 years. In the future, Fisher foresees quality improvement as the greatest opportunity the company has for further growth and increased profitability.

Motorola's push for higher quality has been going on for a decade. It was launched after a single executive, Art Sundry, in the Communications Sector, rose at a management meeting and made an impassioned plea for Motorola to be more responsive to customers and to improve the quality of its products and service. Sundry's plea triggered a plan put together by a group of executives in the Communications Sector in 1981 to make a tenfold improvement in performance in 5 years, based on a set of exacting measurements. To set a benchmark for Motorola's progress in 1985-86, according to Richard C. Buetow, vice president for quality, a group of Motorola executives toured Japanese plants considered the world's best in such consumer products as watches, television sets, VCRs, and calculators. They wanted to find out firsthand why these Japanese companies had earned such a reputation for high quality.

They found the Japanese plants to have process defect rates some 500 to 1,000 times better than ordinary electronics companies: one to two parts per million, Buetow says. Even with robust designs, which could tolerate some inconsistency, the Japanese companies still relentlessly reduced process variations.

Most American executives still believe that quality costs, according to Buetow. But the Japanese factories were putting out the highest quality consumer electronics products in the world, and their costs were lower than their competitors'. This benchmarking encouraged Motorola to redouble its efforts and added to the recognition that product quality is only 5% to 10% of improving a company's performance. Other factors chosen for improvement included quicker response to customers and cycle times for initial designs as well as the order-to-delivery process. Every management meeting now begins with a review of quality performance. From 1986 to 1988, the measurements of defect levels dropped from parts per hundred to parts per million, Buetow says, and six-sigma quality became a global corporate goal.

Motorola recognizes that it can't possibly reach its six-sigma goal without much higher levels of quality in its suppliers. So, after winning the Baldrige Award, Robert Galvin, Motorola's chairman at the time, wrote to all vendors—even banks and insurance companies—asking them to come up with plans to enter the Baldrige competition themselves. At the same time, Motorola offered its assistance to suppliers to help them learn the techniques necessary to reach world-class performance.

～ *Quality Defects Once Cost $800–900 Million a Year*

Quality *does* pay, George Fisher insists. He says calculations indicate that before Motorola raised its quality levels, it spent at least 5% to 10%, and in some cases even as much as 20%, of its sales dollars on poor quality. "That means we were wasting at least $800 million to $900 million a year," he says. To cut that huge waste, Motorola adopted a wide range of the tools now being used to boost total quality throughout U.S. industry, including statistical process control, the use of cross-disciplinary teams to reduce cycle times, partnerships with suppliers, and increased training. To emphasize the philosophy of continuous improvement, Motorola set up financial incentive programs with formulas rewarding both groups and individuals within those groups, depending on the measurable improvements they are able to achieve each year.

Perhaps the strongest push, which is also emphasized in the Baldrige criteria, is toward customer satisfaction. "Probably the best route to growth is to really listen to customers," says Fisher, "That can help build market share." Fisher and other executives, down to factory managers and design engineers, now make a practice of visiting customers not only to identify problems but also to seek market opportunities. That's necessary, Fisher explains, because sometimes the customer may not even recognize opportunities for improved products or services. Although some customers get right on the line with any problems, or with complaints and suggestions, others may not

~ MOTOROLA SEEKS ELUSIVE SIX-SIGMA TARGET

To get dramatic results, Motorola executives agreed that they needed to strive for very tough goals. When they fixed on a goal they dubbed *six sigma* as their 1992 target for improving operations and products, even the most avid proponents of quality were impressed.

What does six-sigma mean? In short, it means that Motorola wants to to be free of errors and defects 99.9997% of the time in *all* that it does. That means fewer than 3.4 defects per million units.

The name of the effort comes from statistics. *Sigma* is a statistical measure of variability around an average. In a normal distribution, represented by the familiar bell-shaped Gaussian curve, the values falling within one standard deviation, or one sigma, of the mean, or average, value, include 68.27% of the total. Move out three standard deviations, or three sigma, to either side and about 97% are included. Six-sigma covers 99.9997% of the distribution under the curve.

Take the example of circuit boards coming off a production line. Shooting for six-sigma means Motorola wants 99.9997% of the boards to perform within specifications. Consider a 10,000-ohm resistor that goes onto the board. One way to keep the board's performance within the six-sigma range would be to design it so that an individual board would still work within specs even if the resistor happened to be far from the mean. By allowing more variation around the average value of 10,000 ohms, say plus or minus 2,000 ohms, the goal of 99.9997% of the boards free of defects might be more easily achieved. That is what engineers mean by a *robust design.*

Another approach, of course, is to keep the process variation within very narrow values through tight control. In other words, the variation of the resistors themselves would be kept within a very tight range, say plus or minus 100 ohms. Either way, reaching six-sigma either side of the mean picks up all but the tips of the tails of the distribution curve to include 99.9997% of all values.

---

find it easy to articulate just what is needed or what may be causing difficulties. Those silent customers too often solve their problems by finding a new supplier.

Along with the six-sigma goal, Fisher cited five other initiatives being pursued at Motorola. They are

- an effort to be an international leader in cycle time management, starting with product design and going through to delivery to customers once an order is received
- manufacturing and technological leadership in all product sectors
- participatory management within all its groups around the world
- cooperative management across departmental and divisional lines
- profit improvement, which Motorola expects as a result of all of the foregoing

Fisher points out what he considers a major weakness of U.S. industry that Motorola is striving to overcome: the organization chart. "Organizations are *not* built to serve customers, they are built to preserve internal order. To customers, the internal structure may not only mean very little, it may serve as a barrier. Organization charts are vertical, and serving customers is horizontal," he believes. Motorola's answer is a participatory management approach, aimed at allowing each individual in the company to compete to the limit of her or his ability. "We don't want to treat people like robots. We want to maximize human talent," explains Fisher. Groups must listen to each other and work together to serve customers. Participatory management systems are basically performance-based, he says, combining the efforts of both the group and the individual.

These are pretty heady ideas, even revolutionary to much of corporate America. So it is important to see how well they are propagating out across Motorola's organization, with about 100,000 employees all over the globe.

Before participatory management can work, it is essential that the work force have the skills needed to grasp the fundamentals of manufacturing processes and other parts of the business. Therefore, the first problem faced in changing the business culture of such a far-flung enterprise is to get everyone to speak the same language. In the United States, that means English. Globally it means that all Motorola managers and employees must interpret terms similarly even when translated into different languages. A unified culture makes training much more effective and speeds change across the organization, according to Fisher. But the company faces some formidable problems in reaching its tough goals.

"In Schaumburg alone," says A. William Wiggenhorn, corporate vice president and director of training and education, "there are 44 different languages represented in the school system, and the parents of those kids work in our plants." Some 1,200 workers already have taken voluntary courses in English, and the company plans to make this mandatory for non-English-speaking employees. Ninth-grade math skills are required, at a minimum, to plot and interpret the data needed for statistical quality control, yet many American workers are far below this level, according to Wiggenhorn.

～ Japan's High Schools Teach Statistical Methods

"In Japan they teach statistical quality control in high school," Wiggenhorn points out. Thus, it costs about 47 cents per worker for automaker Toyota Motor Corporation to teach new people its statistical quality control methods. In the United States, Wiggenhorn estimates, teaching statistical quality control costs about $200 for a worker with ninth-grade math capabilities, and $1,000 for one who is subpar in math. Consequently, Motorola executives have become much more involved in helping improve local schools. The company already has close links to universities, and a new executive has been added to coordinate education below the university level.

Training methods also can be improved, he believes. Some people learn much better through hands-on experience than from lectures. Yet the common approach in the United States is to give generalized instruction. In Japan, by contrast, Wiggenhorn finds that training is linked directly to applications on the job. After instruction, the Japanese worker has to show in the workplace what was learned. Many U.S. firms, especially under affirmative action programs, are afraid to test workers for fear of legal problems, Wiggenhorn believes. That's because the testing philosophy is wrong, he feels. "It's the 'I gotcha!' approach rather than helpful feedback," Wiggenhorn says.

Measurements are an important phase of all Motorola's programs. They are used to check progress toward very tough goals. Along with such factors as product and process quality, cycle times, and customer satisfaction, Motorola added a category it calls *mindware,* to monitor the skill levels of its work force in such areas as language, math, problem solving, and teamwork. The company has already established baseline levels. One goal, says Wiggenhorn, is that each employee will get a minimum of 40 hours of training, and the results will be carefully tracked.

Cross-disciplinary teams, with members plucked from disparate business units, often go through training together, with the focus on serving customers. Ways to increase cooperation among groups were added to the training agenda after Motorola ran an experiment by swapping teams between its semiconductor plants in Phoenix and Izu, Japan. The U.S. teams proved to be much more competitive with one another, while the Japanese teams cooperated more, such as by keeping a log on each shift and passing it on to the next shift. Methods for cooperating, rather than competing, with other groups have since been added to training. When setting up a Motorola Design Center for the automobile industry in Detroit, cross-functional training proved valuable in linking the company's semiconductor and automotive groups, for example.

A contract book for new projects is another mechanism Motorola is using to engender more cooperation. This book documents all phases of design of both the product and manufacturing processes, including hardware and software, as well as other factors such as methods for order processing, service, and maintenance. Managers of each group must approve their portion of the plan, committing to six-sigma quality. These plans set very tough goals.

~ *The Bandit Factory Is Quick on the Draw*

Gordon J. Comerford, corporate vice president and director for the Communications Sector, cites the case of a portable radio that normally took 10 to 12 weeks from the time an order was received to shipping date. "It now takes 4 days," says Comerford. Even that wasn't good enough for the pagers produced in Motorola's state-of-the-art factory in Boynton Beach, Florida. The time from when an order for a pager with a particular set of options goes into the computer from any sales office until the time the unit comes off the line is now down to 72 minutes.

The Boynton Beach pager plant was known internally by the code name Operation Bandit, because it was put together by abandoning the traditional American "Not Invented Here" syndrome in which everything is designed from scratch. Good ideas were borrowed from wherever they could be found, including Xerox Corporation, the Japanese, and L.L. Bean (the Maine-based catalog marketer of sporting goods), as well as from other Motorola units, according to quality vice president Buetow. Then the design team added its own improvements. "We had to borrow ideas," explains Comerford, "because when we looked at the goals we had for this plant there was no way the 22 or 23 engineers putting it together could have designed it from the ground up on their own within a few months."

Change in a corporate culture like that which has occurred at Motorola can't happen without internal tensions. What is helping the company cope with the friction is a willingness to face such problems head-on and to encourage communication to deal with setbacks. One example is the bonus system, which provides financial incentives to boost quality. Every 2 years, each business unit within the company is audited by quality specialists from other units in a sort of mini-Baldrige exercise. The quality specialists review the findings with managers of the units so that failings can be addressed. All units in the company are ranked by numerical scores.

In groups that meet especially tough goals, financial incentives have ranged up to 25%, or even 40%, of salary in rare cases, Comerford says. Functional groups have bickered about whether the improvements measured were equally difficult, between sales and factory work, for example. Initially, the focus was heavily on bonus amounts rather than the underlying goals of the program.

Rather than killing the program, as some companies faced with such internal squabbling might have done, Motorola took steps to communicate more precisely how the incentives work. In a key step, Motorola issued a new management manual that clearly lays out in simple language and with specific, real examples how the process can be more effective for both the individual and the company.

∽ *The Pitfalls of Flattening*

Leveling the organization chart is another process fraught with pitfalls, explains Rich Chandler, vice president and director of manufacturing for the Radio-Telephone Systems Group. In his Schaumburg plant, which produces base station and other equipment for mobile and cellular phone systems, the organization has been scaled down from seven levels to four—in some cases even three—in recent years.

Flattening the organization requires radical changes in job definitions and in the way things are done, Chandler says. In a plant that puts out 17,000 circuit boards a week of some 350 different types with up to eight-layer complexity, communications must be superb. It has taken nearly 2 years, and the cooperation of a local school, just to get everyone speaking English. New hires are tested for basic language and math skills.

Chandler is an impassioned advocate of participatory management. In fact, he defines a manager as anyone who utilizes company assets, whether it be a business unit with $1 billion in sales and 1,000 employees, a $500,000 tester, or a soldering iron. "That makes everyone a manager," he says. "Too many American companies think that everything revolves around sales management, and that the guy on the shipping dock doesn't matter." In his organization there is only one grade of worker—manufacturing. Pay at Motorola, a nonunion company, is based on how many skills an employee has mastered rather than just seniority.

Chandler's goal is what he calls a self-directed work force, with maximum flexibility because of extensive cross-training. Building communication equipment is a batch operation, and it is clear that different work groups have organized their work spaces and materials flow to suit their own needs. Chandler explains that this freedom has led to at least five different materials systems in the factory. Management lays out goals but does not impose details for how each job will be done. Workers manage their own planning, scheduling, maintenance, and testing, and they have the right to shut down and fix the line when something is wrong.

What might seem like chaos to a more traditional factory manager has resulted in circuit boards moving from start to finish in 2 days, on the average, compared with 12 to 15 days in 1987, according to Chandler. The plant's target is 1 day.

Each group also is challenged to cut defects in half each year, and charts on the factory wall show the progress being made.

A tough-minded executive might ask, what's the bottom line for this headlong quality push? Afer winning the Baldrige Award, Motorola's sales climbed at a rate greater than 15%, while profits moved up even faster. Maybe that's why companies from all over have flocked to Motorola to learn how they're doing it.

∽ MOTOROLA WORKERS LEARN BY RUNNING A TABLE-TOP FACTORY

Many people learn much more from hands-on experience than from lectures. So what better way for workers to learn about manufacturing than to give each a shot at running a factory? To A. William Wiggenhorn, corporate vice president and director of training and education at Motorola, that's the idea behind the company's Technology Awareness Laboratory. Opened in March 1989 at the campus-like corporate headquarters in Schaumburg, Illinois, the lab features a novel, robotic table-top factory.

Workers use computers to automate the collection of machines, which takes up no more space than three desks, to produce prototypes of their own plastic "products." By mid-1989 some 170 factory supervisors and line operators had gotten at least 8 hours of training in the lab. No computer experience is required, and according to lab director Art Paton, a few students could not even read. Yet by the time they finished the lab program, all of them were familiar with basic terms and concepts behind factory automation. Understanding the technology makes workers more productive, says Paton, and helps them to suggest improvements. This training also is helping Motorola achieve its daunting goal of six-sigma quality.

Before they take a crack at running the little "factory," each student at the lab spends time at a workstation equipped with a desk-top computer, a programmable robot arm, proximity and optical sensors, a switch and a machine vision system with a monitor to show what the system "sees." In a simple exercise, the robot is instructed to get ready for programming through a command from the computer.

The objective is to move an object from one point to another using point-to-point control. Joysticks control the linear and rotary motion of the robot's shoulder, elbow, and hand. The operator manually positions the robot over the object, picks it up, moves to the new position, and so on, until a sequence of operations has been programmed.

While programming the robot, the operator sees binary signals fed from the sensors, which change as the robot moves, displayed on the personal computer. The rudiments of machine vision and image processing are also demonstrated.

After a series of exercises, the class goes out to the table-top factory, where plastic blanks of different colors are selected, machined, inscribed with labeling entered by computer, and assembled into a small product. The process makes use of robots, sensors, machine vision, and the other tools that the class has just studied. The class sees that standard products can be made three times faster than customized versions.

The lab has big advantages over on-the-job training, according to Paton and Kathy Burgos, who runs the table-top factory. A real plant is noisy and busy and many types of problems don't arise very often in actual production. Also, the line doesn't have to be slowed down or stopped for training. In the lab, workers can run simulations to show the effects of mistakes that might be costly or even damaging in a real factory setting.

"With automation, workers can't see what's going on behind the scenes in the factory. Such things as orders and information flow are invisible," says Paton. "What we show them adds confidence because they learn what's happening behind the scenes. It's like putting a rock in the middle of a stream to help someone across."

Chapter 5 ∽

Xerox Fights Back from the Brink

To an American corporation, regaining markets once they've been lost to Japanese competitors appears about as easy as reassembling Humpty Dumpty. The task is theoretically possible, but succeeding at it might take more men, women, and horsepower than even a king could assemble.

Over the past decade, Xerox Corporation has learned quite a bit about putting things back together. It was during the late 1970s that the Stamford, Connecticut, company was jolted by the fact that it was being soundly thrashed around the world by its Japanese rivals. Xerox went from an 18.5% share of the U.S. copier market in 1979 to a 13% share in 1981 before bottoming out at 10% in 1984—a humiliating turn of events for a company whose name is synonymous with photocopying.

Well, Xerox is back. The company's U.S. market share revived to 12.8% in 1987 and 13.8% of a $13 billion market for copier hardware in 1988, according to Bob Sostilio, associate director for copying and duplicating services at Dataquest in San Jose.

How did Xerox pull it off? It reexamined and redefined its traditional concepts of quality, then set demanding new goals in every one of its operations—design, production, inventory management, supplier relations, and marketing. It also took aim at increasing the efficiency of its corps of middle managers by improving worker relations and enhancing problem-solving abilities. "At this point, I think we've met or exceeded every one of those goals," says James E. Sierk, a former vice president who headed Xerox's quality office in Fairport, New York.

The extra effort paid off in another important way. Xerox's Business Products Division won the coveted Baldrige Award in 1989. Winning the Baldrige competition was no easy task; at one time at least 17 team members were working full time with Sierk, the vice president for quality, to prepare the entry and to gear up for site visits. Even for a giant company such as Xerox, such a commitment would be hard to justify if the only payoff were a trophy and a prize few Americans yet recognize. Indeed, Xerox says the prime motivation for applying for a Baldrige was the process, not the award itself. According to Sierk, the company viewed the Baldrige effort as only one of the means it employs to reach its own definition of quality: "totally meeting our customers' requirements."

"To see things through our customers' eyes—that was our goal in the Baldrige process," says Bob Wagner, manager of public relations programs for marketing and customer operations at Xerox's Rochester, New York, location.

∿ A Frightening Vision in Japan

Xerox began its about-face around 1979 by taking a hard look at the competition, partly through the eyes of longtime affiliate Fuji Xerox Ltd., which commands about 20% of Japan's copier market, according to Sostilio. Former chairman and chief executive David T. Kearns also took an untold number of trips to Japan. The relationship with Fuji Xerox was Xerox's window to Japan, enabling the company to see first-hand what the Japanese were doing on the quality front.

"What we saw was frightening," recalls Sierk. "They simply were eating our lunch. They were getting products into the market faster, their products were more reliable and were being produced at less than half the cost of our machines. In fact, they were selling their machines for less than it cost us to make ours."

There was the usual denial, of course. "We tried to convince ourselves that the numbers were wrong," says Sierk, "or that it was a currency glitch or that the Japanese must be cheating somehow. Finally we realized that they were just managing better than we were." For example, whereas Xerox Corporation routinely dealt with thousands of suppliers, Fuji Xerox bought the same commodities from just a few hundred.

That was the beginning of Xerox's benchmarking process—a procedure by which a company identifies key goals in design, manufacturing, product performance, marketing, and so on, based on its own history or on that of another pace-setting company. In high-end copiers, for instance, Xerox benchmarks arch rival Eastman Kodak Company; in distribution, L.L. Bean, Inc., the Freeport, Maine, outdoor and sporting goods specialist, was tabbed.

Xerox's internal reexamination also led to its Leadership Through Quality process in 1983. Since then, the company reports, some 110,000 Xerox employees worldwide have been inculcated with Leadership Through Quality principles. The company credits former chairman Kearns, through his visits to Japan, with inspiring the program. He was one of the first at Xerox to go through its quality program. Some of the results thus far, according to the company:

- Customer satisfaction has improved by 38%.
- Manufacturing costs have been cut in half, primarily through improved design in the early product-development phase.
- Overall product quality, as measured by defects per 100 machines, has risen by 93%.

One of the first basic changes at Xerox was the recognition that the old ways of doing things just didn't work anymore. Managers throughout the organization cultivated a new awareness of actual customer requirements. Just being able to fix something when it broke was no longer acceptable. As Sierk puts it, "Our policy is that if the customer's satisfied, it's not because he doesn't get a bad product. It's because the bad product never got made in the first place."

The underlying principle here, of course, is quality—even though Sierk admits that the word has different meanings to different people. "I think it means meeting customer requirements," he says. "In Japan, producers and designers are thinking about what customers will require in the future, even though they (the customers) don't even realize it yet."

However quality is defined, most experts understand that it costs when it is missing—that is, the cost of rework, scrap, nonroutine customer service, and so on. At Xerox, it runs as high as 30% of sales, by Sierk's estimate. The best way to pocket those costs, he says, is by minimizing the variability of products, processes, and people in order to design a predictable, more controllable system. Nor are such systems limited to engineering and manufacturing. In the bookkeeping department, Xerox cut the number of errors in supplier payments from one per 100 to one per 1,000. The benchmark target in this case is one error per million or fewer.

Of course, to achieve such a goal a company must identify the root cause of any error. "Say there's a leak in the roof out in the plant, and you've got water on the floor," says Sierk. "You can fix that problem by mopping it up or putting a bucket under it or fixing the roof. But you dig deeper and learn that there's a big tree outside and a branch is rubbing on the roof. You can cut the tree down, but if you set a policy prohibiting trees from within 40 feet of any plant, you've solved that problem forever.

"It's the same thing with customer dissatisfaction," he continues. "Is he dissatisfied because of poor assembly? Substandard parts? Why was it assembled poorly? Why were the parts substandard?"

∿ Starting Point: The Design Process

For any manufacturer, the most logical place to begin implementing a quality policy is in the design process. Xerox was no exception and set out to forge much closer links among design, engineering, and production. "In the old days, the designer would finish a project and just hand it over to the plant manager, in effect saying, 'It's all yours; I'm going to do something else,'" explains Lou Marth, product manager of Xerox's New Build Operations in Webster, New York.

"A lot of us were skeptical when the Leadership Through Quality program came in," says Thomas Mooney, manager of advanced products technology and engineering. "We called it the flavor of the month. But we've seen much closer ties among design, engineering, and marketing, and much more discussion of customer satisfaction at all levels."

Today, the chief engineer sits astride the whole process by assuming responsibility for virtually every phase of product development and delivery, including capital tooling and early manufacturing processes. Engineers are required to adopt regional districts and visit customers regularly. In addition, every new product-development team includes a marketing manager who can relay customer requirements and expectations to the designers.

In addition to improving plant operations, such a strategy has cut the company's customer-service costs. "If you can design a machine that's easy to assemble, it will also be easy to service," says Marth. "That has a big impact on the profitability of our maintenance agreements—service costs can really break your back."

Some 1,700 employees at the New Build Operations turn out 10 copier models. Among the machines is the 5090 line, part of the 50 Series of copiers introduced in 1989 as part of the 50th anniversary of xerography; the 50 Series now consists of nine models. The New Build employees are divided into several business area work groups, or BAWGs. At present, the ratio of supervisors to workers runs about 1 to 40. Marth notes that the company hopes to cut that ratio in the future by granting more authority and responsibility to the work groups.

At one time, recalls Marth, "New Build Operations wasn't very well organized. Parts would move all over the building before they reached their destinations." The facility was reconfigured to minimize the traveling distances of each part or assembly and to maximize the operation's overall efficiency. Assemblers' ideas and opinions were widely used in the redesign, and workers are still encouraged to suggest new ways to cut unnecessary steps and smooth out the parts delivery and handling systems.

Although all copier models are made in separate areas of the New Build Operations, designers are urged to use common parts and generic assemblies wherever possible to reduce the total number of components and materials on the plant floor at any given time.

⌇ Quality on the Factory Floor

In Xerox's manufacturing areas, several unusual features are noticeable. One is that any worker can stop the conveyer that carries the assemblies from workstation to workstation whenever he or she spots a problem—such as a misaligned or improperly connected component—and discuss the problem with others on the line. Marth notes that such events are relatively rare, however, since most parts are designed to be failsafe on the assembly line. It's just about impossible to misassemble a component because of its geometry.

Another unusual feature is the daily computerized defect report, posted at each work group station, which identifies defects that have been spotted, the stations at which each defect occurred and was found, and the number of times that defect has occurred during a given time period. Work group members meet formally 1 hour a week to review and resolve these or any other assembly problems.

At some assembly stations, workers use scanners to signal when they're running out of parts. The signal is picked up by a small computer aboard one of the lift trucks that roam the floor. Once the signal is received, the operator retrieves the needed components and delivers them to gravity-fed bins from alleys behind the workstation. Thus, the delivery of parts does not interfere with the assembly process.

Not surprisingly, the company's revised design and manufacturing procedures meant a radical overhaul of inventory management procedures and supplier relations. "We saw that we had millions of dollars in unused inventory," says Marth, "which tied up a lot of our assets." The result was a just-in-time (JIT) program—an admittedly difficult concept to put into practice because of Xerox's status as a worldwide procurer. Nevertheless, says Sierk, the program has paid off handsomely. While the company maintained a 3-month inventory during the early 1980s, stocks are now down to about 40 days' worth and heading for 20 days. Whereas Xerox routinely inspected 80% of incoming shipments 10 years ago, less than 15% (primarily new parts or materials) are

∿ Xerox Managers Now Communicate in a Common Language

There's no doubt that Xerox's Leadership Through Quality process has had a powerful, positive impact on the company's performance during the past 5 years or so. But the program goes far beyond market shares and net sales. As the formal concepts of total quality took root throughout Xerox, workers at the middle-management level were often startled at the higher efficiencies, improved worker relations, and enhanced problem-solving abilities.

One key reason was that workers at all levels could finally communicate in a common language, explains Lou Marth, product manager of Xerox's New Build Operations. When the Leadership Through Quality program was launched, he says, employees were divided into "family groups." The first group (which included former CEO David T. Kearns and other top-level executives) underwent formal quality-training sessions—designed by Philip Crosby and other consultants—of up to a week. This core group then became quality-assurance teachers to the next lowest family group, and so on down the hierarchy. At the end of 18 months or so, everyone in the company had spent at least 20 hours learning the same tools, methods, and terminology of quality assurance.

"Previously, everyone had their own ways of solving problems," says Marth. "We were all speaking different languages. It also was hard to keep personalities out of these tasks—problems were identified by saying something like 'I'm having trouble with so-and-so over in production,' or some such thing, and so-and-so in production would reply, 'I'm not the problem; it's Joe Blow in design,' and so forth.

"Now we have specific step-by-step methods, such as benchmarking, that identify and state the problem explicitly and provide means of solving it." One result of this structured approach, Marth adds, is that "I've become much more aware of what customers and other employees need and want from me. I try to find out what they need rather than making assumptions that might not be true."

To be sure, middle management was skeptical at first and somewhat fearful of giving up power as lower level workers assumed more responsibilities. There was also a natural resistance to change in the way jobs were to be performed. "But most of us knew that we had to change in order to compete," says Marth. "As time went on, most managers took a positive view—we saw that delegating authority to people further down the line really does work."

---

checked today. "That also means we need only about one-eighth the number of inspectors we used to have," says Sierk.

The obvious implication of that policy is that Xerox demands exceptionally and consistently high performance from every one of its suppliers. Under its Centralized Commodity Management program, Xerox slashed the worldwide number of production vendors from 5,000 in 1982 to just 400 in 1990. "If that was to pay off," says Marth, "we had to make it attractive for the best suppliers to invest in training and tooling. We wanted them to know that we wanted a long-term relationship."

Apparently, the 400 survivors got the message: reject rates of purchased parts and materials plummeted from about 30,000 per million parts during the late 1970s to some 300 per million today, according to Robert F. Willard, manager of materials operations. The company's goal is approximately 100 per million.

"By creating a mutual dependency, we were able to get rid of the old antagonistic, arm's length ways of doing business," explains Marth. "Like most companies, we had always spread our business over several suppliers of each commodity. What that told them was that we didn't trust them and that competition will force them to lower their prices. All of those principles are fundamentally wrong. It makes much more sense to assure each of them enough steady business that they can afford to work with us in design, the JIT program, statistical process control, and so on."

To accomplish those goals, Xerox set out to get itself on every supplier's list of top five customers; in return, Xerox demanded that each supplier guarantee that once a part is approved, it will be delivered on time and within the tight performance specifications that make inspection unnecessary. That eliminates not only the need to send out for new bids every year, says Marth, but "it saves the vendor a lot of effort and time. Now, his salespeople don't waste their energies sitting in our lobby waiting to talk to our buyers."

Not surprisingly, many suppliers balked. "Some of them didn't take us seriously, and they were cut," recalls Sierk. "Now we have vendors that are working very hard to get back on our list." That won't be easy, and no one gets on the list through cold calls, warns Willard. Although Xerox adds about 10 new suppliers a year worldwide, Sierk says the emphasis is on vendors who can bring new technologies, early-design capabilities, or very strong customer support to the party.

Mooney recalls that in 1987, Xerox was working with an Alabama manufacturing company. "They came up with a new plastic molding process which saved 40% on the price of a part. That company is now an approved second source for that part."

Other vendors agree that Xerox sets rigid standards, but not many complain. "They have very high expectations of their suppliers," says Marie Gay, director of corporate marketing for Galileo Electro-Optics Corporation in Sturbridge, Massachusetts, a Xerox supplier since 1977. "They tell you up front that you must produce perfect parts and how they'll work with you to reach that goal. If you're not ready for that commitment, you're wasting your time there." Galileo supplies a glass-coated wire assembly, called a dicorotron, used as an electrostatic charging device for Xerox's Marathon series of copiers.

In most cases, Xerox's high demands pay off in its suppliers' own operations. Gay notes that in order to meet Xerox's expectations, Galileo set itself a goal of zero defects—not just for Xerox, but for all its customers. During Galileo's first year as a supplier, she says, only 14 of the first million components had defects. In 1989, Galileo won Xerox's Award of Excellence for the fourth year in a row. "They tell us that we've now gone three years running without a single defect. That makes us feel very special."

∽ 60,000 Questionnaires Tell the Story

The ultimate goal, of course, is customer satisfaction. Xerox maintains several programs to track that parameter and to respond to changing customer demands.

One example is the 8-year-old customer satisfaction measurement system—a monthly survey comprising 60,000 mailed questionnaires. Under that system, says Sierk, every customer will have the opportunity to grade the company at least once a year on such areas as equipment performance, sales, and service. The results are tabulated and used to introduce necessary design or operating changes. One example of such a change came from a field test on a Xerox typewriter. "We found that a lot of operators didn't like the feel of the keys," recalls Marth. "We redesigned that, and it was a landmark in increasing the market acceptance of that machine."

Other programs include the Early Warning System, which monitors customer acceptance of all new Xerox products during the first 9 months of launching; the Faultless Install System, which assesses the company's installation performance; and the Post Installation Survey, which measures customer satisfaction with the entire delivery system.

Changes also can come about internally. Marth recalls that the company modified the internal software on one of its copiers so that it would check its memory more quickly after being unused for a long period of time. "We didn't have to do that," says Marth. "There weren't very many complaints but we thought it was a smart thing to do."

Xerox has clearly come a long way after its painful lesson 10 years ago, but don't expect to see any letup in its obsession with quality. If the company has learned anything in the past decade, it's that quality is a slippery commodity, subject to ever-changing customer demands and expectations. It has also undoubtedly learned that the quest for quality isn't a program—it's a process that never stops. Says Marth: "I tell prospective employees that if they don't like change, they shouldn't come to work at Xerox."

Chapter 6 ⁓

A Look at Some Other Baldrige Winners

Any company hoping to win a Malcolm Baldrige National Quality Award has a growing number of models to study: the list of past winners. Each company that enters the competition must agree that if it wins, it will share information about its total quality methods with other U.S. businesses. Baldrige victors report that they are besieged by thousands of such requests. How well have they responded? Fortunately, these outstanding companies have proven to be generous with their time and ideas, although in some cases information seekers must fill out request forms and tailor visits to a limited selection of dates and times.

Any concerns about the Baldrige judging process favoring a particular type of industry or business have also been allayed. It is not surprising that a number of winners, such as Motorola, Xerox, and IBM have been in high technology sectors where global competition is particularly intense. The variety of other enterprises that have either won an award or become strong contenders, however, suggests that the quest for total quality can be successful no matter what the activity or business sector of a particular organization.

Winners from each of the first years of the competition not only reflect this diversity but also suggest that focusing on improving total quality can be an effective mechanism for success in the face of adversity in any marketplace. The nuclear power industry, for example, has been beseiged in recent years because of concerns over safety and the environment. Yet in 1988, the first year the Baldrige Awards were presented, one of the winners was Westinghouse's Commercial Nuclear Fuel Division. It is an encouraging sign to the nation that a firm with a mission of such intense public concern is dedicated to total quality concepts.

Another factor, however, drove Westinghouse's quest for total quality: increased foreign competition. Winners in each of the following two years came from sectors facing even more intense global competition than the nuclear power industry: textiles and automobiles. In 1989, Milliken & Company, a textile producer, won the ward. In 1990, the beleaguered U.S. automobile industry produced a Baldrige winner: the Cadillac Motor Car Division of General Motors. The experiences of these three

47

organizations in adopting total quality as a means for meeting tough marketplace challenges teaches lessons useful for any type of enterprise.

~ Westinghouse Nuclear Fuel: Using Technology to Improve Quality

Through the 1970s and early 1980s, regulatory barriers and greatly increased costs slowed construction of planned nuclear power plants and kept new plans off the drawing boards. As a result, the nuclear fuel market shifted from a primary emphasis on supplying new nuclear plants to maintaining and reloading fuel at existing plants. In addition, international competition increased in the U.S. marketplace.

Up to that time, Westinghouse's Commercial Nuclear Fuel Division had taken pride in meticulously satisfying existing regulations. It had managed a program of product inspection and material traceability that was considered a model in the industry, while following required procedures and producing copious documentation. The division knew that it had built a reputation for excellence because its performance, like that of its competitors, was constantly being audited by customers.

In spite of this outstanding record, however, management of the Nuclear Fuel Division recognized that even more needed to be done to make the division stand out in an increasingly competitive marketplace. Rather than simply following a passive, prescribed regimen based on rules and procedures, a more proactive approach, based on long-term continuous improvement toward total quality, became the division's new main objective.

The corporate level provided important help in shifting direction. Westinghouse had determined that its various business sectors needed strong support to meet the growing challenges of maturing markets and increased global competition. To help nurture these competitive efforts, Westinghouse was the first Fortune 500 corporation in the United States to set up a productivity and quality center. The staff of some 125 quality specialists at this multimillion dollar corporate center developed a universal management model for a total quality process, which could then serve as a blueprint for individual programs within each division.

There are four essential elements in this total quality model:

- management leadership
- product and process leadership
- human resource excellence
- customer orientation

The process must begin with management creating a culture suitable for nurturing quality efforts at all levels. This must be followed by planning, communications, and accountability, all aimed at continuous improvement toward total quality.

The Nuclear Fuel Division began to apply this corporate vision to its own operations in the early 1980s. Each year, a new major quality objective was developed as a focal point for the division's journey toward total quality. In 1984, it set out to become recognized as the highest quality supplier of commercial nuclear fuel in the world. Achieving such a lofty objective, management realized, would require the transformation of the working environment to a customer-driven, quality-obsessed culture.

Rather than turning this mission over to an individual quality executive, the company set up a quality council made up operating managers. This sent a clear message to employees that total quality concepts needed to be applied in everyday working situations. The council's duties include setting annual goals for the Nuclear Fuel Division and then monitoring and reporting on performance. Each year the council, with active participation by workers, develops a quality plan, which is distributed to all employees.

To gauge progress toward total quality, hundreds of measurements were developed within the division. Eight key factors, called pulse points, serve as indicators of overall success. Some of these keep tabs on how well the organization is doing in satisfying customers, including measures of fuel reliability, on-time deliveries of hardware and software, software errors, and even a composite customer satisfaction index. Other pulse points are internal quality measures, such as first-time-through yields of manufactured components and total quality costs. Scores for these pulse points are posted at all division locations to keep employees informed of progress.

Pulse points are only a small part of an extensive communications program that keeps information flowing from customers as well as up and down the organization. Internal recognition programs, including celebrations and parties based on achievements, are common. The general manager gives about 40 presentations a year to provide personal interchanges with some 2,400 employees in four separate locations. Management also fosters an open-door, participative style to keep information flowing freely in all directions. Training is emphasized with more than 1,800 employees participating in quality-related training sessions in any year. Managers also visit customer utilities, and customers are encouraged to visit the division's facilities. They also serve on a fuel users group and on joint quality teams. Employee suggestions and feedback are encouraged, while internal videotapes, newsletters, electronic lobby monitors, and computer messages all help to exchange ideas.

These efforts have paid off. The Division reports that over a 3 year period employee suggestions steadily increased, from 425 in 1985 to 2,000 in 1988, resulting in more than 2,300 quality improvements in manufacturing, engineering, and administration over this period. Corporate awards and honors within Westinghouse have also gone to thousands of teams, many of them from the Commercial Nuclear Fuel Division, and to the Division itself.

The division has applied technology in numerous ways to increase total quality. To better communicate with customers, Westinghouse connected some to its electronic mail network. The division also began to experiment with providing direct access to scheduling and manufacturing status information at the nuclear fuel plant in Columbia, South Carolina, via PC-based networks. The customers could thus keep tabs on their own orders and plan audit visits to division facilities at opportune times. When examining fuel at customers' plants, division employees used photophones to transmit detailed video images of fuel back to headquarters in Pittsburgh.

Fuel rods are produced in the Columbia plant using a highly automated, dedicated process line. A single operator at a control center can monitor the entire process, including fabrication of uranium fuel pellets, loading the pellets into fuel rods, and welding the rod end caps. Statistical process control methods have been invaluable in steadily reducing variability for each step.

Artificial intelligence and expert systems technology also help in reducing variability in manufacturing processes for such critical components as tubing and uranium pellets. By using inductive learning principles, these systems actually get "smarter" the longer they are used. The first year such an expert system was used, pellet yields increased from 88% to 93%, according to Westinghouse. A similar combination expert–inductive learning system was later installed in the division's specialty metals plant.

The results of the continuous improvement process over time have been dramatic. For example, through the use of cycle-reduction techniques, and without side-stepping the exhaustive procedures and documentation required in the nuclear industry, the division was able to cut the product development time for new fuel assembly designs from 7 to 2 years. During the design process, any calculations must be independently verified by a second engineer, and any deviation in production methods must be approved by a design group.

Feedback from field experience and customers also contributes to the improvement process. Some years ago when it determined that the failure of some fuel was due to excessive hydrogen, the division took steps to cut down hydrogen levels in uranium pellets. By continually pursuing this goal, the division believes the hydrogen content in the pellets it produces have become the lowest in the world.

The vital pulse points monitoring overall progress have also tracked striking improvements. The primary measure of fuel rod reliability shrunk by a factor of 10, to less than 100 known to be leaking out of 2.5 million installed, over the period of a few years. First-time-through yields of completed fuel assemblies at the plant in Columbia went up 48% in 4 years, achieving levels in 1988 that hadn't been expected to be reached until the 1990s. In 1988, the division achieved a record of 42 consecutive months of on-time deliveries of all finished assemblies and related hardware, and it marked similar results in software deliveries, including proposals, reports, and documentation.

Total quality costs, including those for internal and external failures, along with prevention and appraisals, went down 30% in 4 years, due mainly to reduced rework and scrap costs. Customer satisfaction ratings, a relatively new measure, went up 6% in a year.

Even more indicative of customer response was a record-setting pace for new business, in large part because of upgrades of existing installations. Customers also cooperate in new product development efforts. The Commercial Nuclear Fuel Division has been selling more of its products to other fuel vendors while increasing its international business in the face of tough foreign competition.

Total quality requires a never-ending quest for continuous improvement. To emphasize this point, in 1988, the year the Westinghouse Commercial Nuclear Fuels Division won the Baldrige Award, it set the vital pulse points to a new baseline requiring even more strenuous efforts to achieve still further gains.

～ Milliken: Learning to Listen

Roger Milliken, chairman and CEO of Milliken & Company, a major textile and chemical firm, traces the beginning of the company's quest for total quality to a management meeting held about a decade ago. One manager told him, "There are only five managers in this room who know how to listen." At the time, there were more than 400 managers present! Milliken recognized the validity of the charge and decided that something needed to be done about it. After some practice, at the end of the conference Milliken stood up on a banquet chair and, raising his right arm, asked all the assembled executives to repeat after him: "I will listen. I will not shoot the messenger. I recognize that management is the problem."

From this beginning developed a new approach to quality in the company, based on leadership through listening and coaching. Dedication to a process of what the company calls total quality improvement led, in 1989, to Milliken winning a national Baldrige Award. A visit to the firm's Spartanburg, South Carolina, headquarters, reveals the extent to which the total quality culture permeates its operations. On the walls of a large room near the main entrance are color photographs and identifications of hundreds of "associates," the term used for every one of more than 14,000 employees.

In the early 1980's when the company launched what was first called a cause removal program, later renamed an opportunity for improvement program, the U.S. textile industry was marked by major layoffs and plant shutdowns due to sharply increased foreign competition. Milliken was facing tough times in spite of being recognized for its use of state-of-the-art technology. After visiting facilities around the world, especially in Japan, executives concluded that the reason for the success of offshore competitors could not be attributed to technological advantages. It was clear that some Japanese competitors with less advanced technology than Milliken were achieving higher quality, less waste, and higher productivity, all with fewer customer complaints. Their advantage resulted from better management methods and personnel policies, leading them to make better use of available technology, the executives concluded.

Top management recognized that Milliken had been managing rather than leading people, according to Steven Gentry, who heads what is now called the Pursuit of Excellence program. "Managing is telling someone what to do, and then checking to see what they are doing *wrong*," explains Gentry, while "leadership is telling people what to do, and then checking to see what they're doing *right*—making them heroes."

Teamwork by self-managed teams was a key element of the new approach. Associates in such teams operate with considerable autonomy and authority. In production, for example, self-managed teams can schedule work, establish individual performance goals, and set up training. Any associate has the right to stop the line because of a quality or safety problem. Cross-training adds flexibility and increases the productivity of the teams. In 1988, Milliken spent about $1,300 per associate on training and also offered training sessions to suppliers and customers.

Because fewer managers are required under this flatter organizational structure, more than 700 managers have been freed to become process improvement specialists since 1981. Extensive data collection on manufacturing and other processes helps to track variation, identify root causes, and ensure that corrective action has the desired results. Real-time monitoring provides data for analysis, in some cases with computer-based expert systems. Improvement specialists work in many sectors in addition to manufacturing, such as billing and customer service. One result has been a 60% reduction since 1981 in the costs of nonconformance to customer requirements, such as discounts for quality problems and freight payments for returned goods.

What has become known as the Milliken quality process involves a variety of different types of teams. In 1988 alone, some 1,600 corrective action teams were set up to address specific business or manufacturing areas. Suppliers have been brought into some 200 supplier action teams, and nearly 500 customer action teams work on better meeting the needs of customers, including development of new products. These teams go through a four-step sequence of defining problems, developing solutions and setting goals, defining a time frame, and then regularly reporting on progress. Results of measurements are displayed prominently so that everyone can see the progress toward targets. Thomas Malone, Milliken's president and a former football player, points out that individuals don't win football games; teams do. The players require coaching, yard markers, and a scoreboard. Cross-training for operating teams gives everyone a better understanding of an entire process, so each individual has a sense of how his or her contribution fits into the team effort.

A standard approach exists for satisfying the needs of Milliken customers, whether internal ones within the company or outside buyers. It starts with asking: "What's most important to you?" Outside customers are also asked: "How do we rate?" and "How do we compare with our competitors?" Surveys help to get feedback from a broad range of customers, but executives also make frequent visits to customer sites. At each monthly policy meeting, for example, each executive must describe a customer visit, explaining what he or she learned as a result.

Benchmarking of the products and services of about 400 competitors is also a regular process at Milliken, not just for comparing performance but also to seek out market opportunities. In 1984, as a result of seeking out such information, Milliken learned that it trailed some of its competitors in meeting delivery dates. To remedy this problem, the record for on-time deliveries was improved from 75% in 1984 to 99%— the best in the industry—by 1988.

One part of the opportunity for improvement program that did not work well was the solicitation of suggestions from associates. Roger Milliken reports that in the first year of the program an average of only one-half suggestion per associate was achieved. One step the company took was to adopt a "24-72 rule." Any suggestion receives an acknowledgment within 24 hours, and, if it is found worthwhile, an action plan is developed within 72 hours. Another idea generated by an associate is a "checkbook" for suggestions, with stubs for record keeping. Some groups within the company offer an award of a restaurant meal after, say, 50 suggestions have been submitted. The program has generated a wide range of ideas across all of Milliken's operation. Something as simple as rearranging an office to make files more accessible to those who use them most, for example, can be put into action immediately without waiting for approval. A more complex suggestion, such as a way to decorate carpeting by spraying colors through dye-injection nozzles, may require extensive research at the company's Research Electronics Laboratory. Recognition at banquets, luncheons, and award ceremonies emphasizes the importance the company places on wide participation in improvement efforts. Each year an Academy Awards-like banquet is held in Spartanburg to reward the top 5% of teams in quality progress. A string of 25 limousines takes the winners from headquarters to the auditorium where the awards are given, and the whole event, including professional entertainment, is covered by local television. About 1,200 people attend.

With such steady improvements to the program, suggestions climbed to 19 per associate in 1989, for a total of 288,000 ideas. Milliken says that 88% of them were actually implemented, but he points out that this still falls short of the 30 ideas per employee received by the average Japanese company, and 50–100 for the leading Japanese firms.

Along with the incremental improvements emerging from such a corporatewide program, management needs to provide a clear vision for the future to help give overall direction to improvement efforts. One challenge issued by management in 1989 is called the 10-4 program, in which a goal was set for 10 times improvement in 4 years in key quality measures. A broad long-range strategic plan issued by top management is to move the company from the mass production and inflexible manufacturing of the past toward flexible manufacturing of small lots, using constant innovation, on-time deliveries, and quick response for new products and services to achieve total customer satisfaction.

To achieve such goals, Milliken must often go around its direct customers, such as clothing manufacturers, to get feedback from end consumers and retailers. The company has been a strong force in achieving standardized electronic data interchange (EDI) all through the marketing chain to achieve much swifter and more precise response to changing tastes and needs in the marketplace. John Geizer, director of management information services, for example, works with such retailers as Sears, K-Mart, and Wal-Mart, as well as such apparel makers as Haggar and Levi Strauss, so that the whole marketplace can become information driven. When a laser scanner reads a tag on a garment with a standard 9-digit code, the information gained can be fed all the way back to the factories supplying the fabrics for garment makers, and even to their suppliers. This is helping to break down traditional retailing, in which large numbers of garments are stocked by stores for a whole selling season. Instead, smaller quantities of a wider assortment will be put into retailing outlets, and information fed back through the chain will allow rapid resupply of the types of items that are selling best. Milliken helps support the EDI efforts of smaller suppliers so linkage can be made all the way back to those supplying dyes, thread, and other basic items.

The efforts of Milliken to better meet world-class competition have not only allowed the company to thrive in the U.S. market but also to become a formidible competitor in overseas markets. Milliken is a major supplier of high-quality upholstery, for example, to major Korean and Japanese automobile manufacturers.

~ Cadillac: Cutting the GM Apron Strings

As a result of the oil crunch and the rise of the OPEC cartel in the early 1970s, General Motors management concluded that increasing oil prices and gasoline shortages would continue and possibly worsen. The Cadillac Motor Car Division responded rapidly and dramatically to these perceived threats. It downsized its Cadillac automobiles, and placed a new emphasis on improved fuel economy. As expected, the overall market for large-sized luxury automobiles declined, but Cadillac did even worse than its competitors. Through the early to mid-1980s, Cadillac experienced erosion of its business and market share. By the mid-80s, management recognized that the earlier predictions had been incorrect and that it had to address a number of growing problems.

A major factor was a loss of the Cadillac image of the past among former customers and potential new buyers. Surveys showed that these consumers felt the new downsized models were too small and underpowered, especially as oil prices settled down and gasoline shortages did not recur. There was also an impression that Cadillac models were too similar to other GM products. They no longer sported the distinctive features that in past times had created a highly individualistic Cadillac mystique, differentiating it from all other automobiles. The image of better quality, which had also been a strong selling point for Cadillac, had also declined as surveys showed that other automobiles, especially from those from Japan, had become the quality leaders.

According to John O. Grettenberger, general manager, the transformation of the Cadillac Division began in 1985, with moves toward simultaneous engineering. Up to that time, the tradition had been a top-down management approach. Even as the process that eventually led to a strong push toward total quality began, management did not intend to work on changing the culture "because we didn't know any better," according to Grettenberger.

To integrate diverse efforts to bring a new automobile to market, however, it was necessary to shift toward a teamwork approach. Various functional groups needed to work on different aspects of the total design project simultaneously and cooperatively, rather than serially and in isolation as in the past. This would permit more rapid response to marketplace changes. Just as important, it would foster better matching of targets for performance, styling, and maintenance with the choice of systems, parts, and manufacturing processes. This was essential, according to Cadillac executives, because in the past, as a new design moved step-by-step through various stages, original goals would often be seriously eroded along the way by compromises. The vision of those who launched the project to meet perceived needs of buyers would be unrecognizable in the product that finally emerged from a lengthy sequential process where important decisions were made by specialists unaware of the reasons for some design choices while under pressure to make maximum use of existing machinery and to standardize on parts and subsystems used elsewhere in the GM product line-up.

Cadillac's effort to instill a spirit of teamwork, starting with engineering operations and manufacturing, provided the groundwork for change, Grettenberger feels. It required a new approach that turned the traditional organization chart upside down. For each layer of the division to work effectively, it needed cooperation from the underlying layers, and managers had to provide support for this process.

To develop a vision for Cadillac, the division sought contributions from all employees. To flesh out a mission statement, Cadillac sought feedback from dealers and customers. It became clear that increasing competition demanded that the consumer's view of Cadillac quality had to be raised. Even then it would not be enough to build well-designed automobiles. Customer perceptions began with reliability and extended all the way through the marketing chain to interactions with dealers and service operations. This meant bringing thousands of workers and 1,600 independent dealerships into a quest for total quality.

The process was given a tremendous boost in 1987, suggests Grettenberger, when the Cadillac division was given much greater autonomy, allowing it to move out from under the corporate umbrella. Top managers spent about a year studying methods for making teams function effectively. Teamwork efforts were moved into the factory and across the organization, as well as out into the dealership network. Cadillac instituted a continuous improvement process, based on trust in employees, with the assumption that they would take personal responsibility for their contributions to Cadillac's overall efforts. Management hoped to foster an environment, according to Grettenberger, that would "give every individual the opportunity to add value for the customer."

The relationship between Cadillac field representatives and dealers was shifted from an adversarial one (as it had sometimes been in the past) to one of cooperation and support. Cadillac began to share formerly confidential plans with both dealers and suppliers. Instead of treating dealers as distributors, Cadillac sought dealers' help as it attempted to better focus on the needs of customers.

About three-quarters of Cadillac's simultaneous engineering teams were expanded to include supplier representatives. The division culled its supplier list and insisted that those retained have continuous improvement programs of their own and that they participate in quality training with Cadillac. There was a 90% reduction in transportation suppliers alone, and the company worked with the remaining suppliers to speed deliveries of parts to assembly plants and finished cars to dealers.

Such efforts resulted in dramatic cycle-time reductions. Average response time to a customer's order was cut 47% in 3 years, with the help of a 58% reduction in hours-per-car for manufacturing. The front end of the Eldorado was restyled in 55 weeks, less than half the time previously required, and a year was cut from the development time for the Seville. Engineering changes were reduced 66% in 4 years.

While pressing for shorter development and manufacturing times, however, management also responded to needs or problems expressed by working teams. In February 1987, for example, with a new model due in September, permission was given for some 700 engineers to take 3 days each for training in teamwork methods. In October 1989, simultaneous engineering teams for a new Eldorado expressed concerns that desired quality goals could not be reached in time for a 1991 introduction. As a result, the launching was put off until 1992.

Relations between management and unionized workers also changed. Grettenberger comments that Cadillac had to drop the previous adversarial stance, recognizing that "the competition is the enemy." Unions made concessions on work rules in 1982, while management agreed in 1984 that any productivity gains resulting from improvement efforts would not result in layoffs. To enable both the work force and management to learn more about the economics of the automobile industry, Cadillac set up courses of 1–2 weeks with the cooperation of such universities as MIT and the University of Michigan.

What Grettenberger terms "diagonal slice" meetings were held, in which workers chose the topics to be discussed, and executives tried to answer questions honestly and completely. To help foster the teamwork needed for simultaneous engineering, Cadillac instituted "engineering jeans day," in which engineers worked on the assembly line along with factory workers.

An atmosphere had to be created in which subordinates at any level were not afraid to speak up if they saw something was not right, or if they saw ways things could be improved. Workers had to feel that their inputs would be considered seriously.

Grettenberg says he felt that the cultural change was taking effect when a worker stopped him on his way to teach a leadership class. "Don't you know that you're supposed to dress casual?" she scolded.

Seven people strategy teams have been formed to examine such issues as employee development, environmental and safety concerns, wellness, and education and training. Each employee now receives 40 hours or more of training each year and in 1990 skilled hourly workers received a minimum of 80 hours of formal instruction in such areas as quality improvement, leadership, statistical process control, and health and safety. Career planning for employees is now linked to the division's strategic business plan. Workers and management in Detroit now have access to a complete fitness center. Since the teams were formed, Cadillac reports that employee turnover has been cut in half.

As a result of all these efforts, Cadillac, which is considered the flagship division of General Motors, made a turnaround in the marketplace. By the late 1980s, Cadillac began to gain increased market share in a down market. In J.D. Powers surveys concerning reliability and durability, Cadillac gained first place among domestic automobiles 4 years in a row and moved into the top five among all carmakers in the U.S. market 5 years in a row. Warranty coverage was extended, from 12,000 miles or 1 year in 1988 to 50,000 miles or 4 years, and new service offerings include a unique nationwide roadside service program.

Much of Cadillac's success can be attributed to the new partnership that has developed between the work force and management, say those involved in the turnaround. In a recent survey of the attitude of factory workers about working at Cadillac, Roy S. Roberts, manufacturing manager, reports that one commented that "the first 20 years were awful, but the last 6 were wonderful." When Cadillac entered the Baldrige competition in 1990, the submission was cosigned not just by the division's general manager but also by its top union leader. Although management felt the division had a strong entry, there was great concern about site visits. Baldrige examiners were free to ask random questions about quality to any of thousands of workers at scattered plants and offices. Indeed, during such a visit an examiner did turn to a nearby assembly line worker to ask what he did to satisfy the customer. Before replying, the worker queried: "My internal or external customer?"

It was that kind of evidence of Cadillac's dedication to the transformation to total quality that helped it win the Baldrige Award in 1990.

Chapter 7 ～

Three Small Companies Win Baldrige Awards

Thousands of small companies all over the United States are beginning to grapple with total quality just like their bigger corporate cousins. The main reason is clear: major corporations are narrowing supplier lists and seeking in-depth partnerships with a chosen few. When these giant firms prepare lists of qualification requirements, total quality systems are often at the top of the list. With future orders from their biggest customers on the line, more and more small companies realize that quality spells survival.

Small company quality programs are not as formal as those of the corporate giants, but these firms do have the advantage of being able to move with more agility. A small company can quickly put together interdepartmental teams to focus on areas that need improvement. When problems are identified, it can alter procedures immediately and then fine-tune them through frequent, often casual, interchanges. Face-to-face communication helps the work force to sense top management's support for their efforts and makes it easier to get everyone pitching in to reach tough goals.

Top management commitment is vital for getting any total quality program rolling, but for small companies it's imperative. Getting started takes investment, personnel, and training time, all scarce commodities in smaller firms. Some small-company CEOs may balk at starting a quality program because of the resources required, but their excuses won't cut any ice with major customers. Corporate giants, recognizing that their own quality depends critically on the things they buy as well as what they make themselves, are not just requesting but *demanding* that all suppliers—no matter how small—develop total quality management systems. More and more often they say *"sayonara"* to suppliers that don't get the message.

The pressure on small firms to boost quality is intense. Even start-up companies find that innovation may not be enough; they also must toe the quality line if they hope to make inroads in the marketplace.

Plenty of smaller firms have wholeheartedly joined the quality parade. Enthusiasm and team spirit help them make up for any gaps in their scaled-down quality programs. To most of them, however, winning a Baldrige small-company trophy appears about as easy as hitting the lottery. Look at the record. In the first 4 years of the competition, when eight small-company Baldrige Awards could have been presented, only three firms won (entries in this category are limited to firms with 500 or fewer employees). Those winners, Globe Metallurgical, Inc. in 1988, Wallace Co., Inc. in 1990, and Marlow Industries, Inc. in 1991, set a standard of excellence that will be tough for other small firms to match.

The United States remains a land of dreams. In spite of the overwhelming odds, lots of small firms *are* gearing up to go for the gold. For any small company hoping to boost its quality, it's worthwhile to review the quality achievements of Globe, Wallace, and Marlow Industries to see what it takes to be a small-company Baldrige winner.

⟋ Globe Metallurgical: Setting a Standard for Quality

In the early 1980s, Globe Metallurgical, Inc., with plants in Beverly, Ohio, and Selma, Alabama, faced some tough problems. Foreign imports of silicon metals and fer-rosilicon products were grabbing markets within the steel industry. Since they were the largest market segments for these types of products, the U.S. ferrosilicon industry was steadily shrinking. Globe Metallurgical, the largest remaining U.S. producer, needed a new strategy to counter the growing imports. Although the invaders' prices were low, Globe noted that product quality seldom matched its own. So the company decided to concentrate on market segments where the demand for quality was strongest, particu-larly the chemical industry and foundries, including those serving the large automobile manufacturers. The key question was whether Globe's approach to quality would make it a big winner in those segments. Management recognized that to survive if the compe-tition got even tougher, it would have to achieve world-class quality.

In 1985 Ford Motor Company approached suppliers with a quality certifica-tion program called Q-1. Globe immediately recognized that in spite of its good record as a supplier it could not make the grade. It was apparent, says Kenneth E. Leach, vice president, administration, that Globe's quality system was detection based rather than prevention oriented as required for Q-1 qualification. Globe was achieving quality by testing and then grading materials coming out of generic processes. The better materi-als identified by this selection procedure went to customers with more demanding specifications. There was essentially no quality planning, employees were not involved in improvement efforts, and statistical process control (SPC) was non-existent. Based on this approach, there was no way that Globe could tailor materials to a particular customer's specific needs.

Because Ford planned an extensive audit of suppliers' quality processes, and a score of at least 140 out of 200 would be required to be considered a long-term supplier to the automaker, Globe took quick action. First, managers and supervisors viewed a series of videotaped lectures by W. Edwards Deming, the quality guru. Deming drummed home the value of SPC in first stabilizing and then continuously improving manufacturing processes. His talks also revealed the futility of trying to boost quality with a system like Globe's, geared to reaction rather than prevention.

With so many deficiencies, according to Leach, efforts toward building a total quality system were launched on three fronts:

1. The entire work force was trained in SPC methods with the help of the American Supplier Institute (chosen because of its close association with the auto industry).
2. A quality manual was developed, helping to provide underpinnings for the mechanics of instilling quality processes throughout operations.
3. Education and training in SPC and quality methods were offered to Globe's suppliers.

Quality information centers, featuring the quality manuals, were located in each department. To ensure a consistent approach, all new employees were trained in the manual's methods before beginning to work on the line. The quality information centers also feature quality news and information, such as quality circle schedules, results of improvement efforts, recognition of employee efforts, lists of any upcoming customer audits, and up-to-date control charts tracking key variables. The charts are plotted on hinged panels for easy access, and if any chart shows a variable running out of set limits, hourly workers jot down comments on how the variable was brought back into line.

Even with the steady progress made through such efforts, Globe began to run into other troubles in meeting the quality demands of customers. Buyers wanted process statistics relating to the particular batches they purchased, and the scores resulting from the old way of grading the output of generic processes were too low. What Globe needed was a way to tailor batches to the specifications of each customer.

∼ *Spending $60,000 on Computers Saves $600,000 a Year*

To tailor batches to the specifications of each customer, Globe installed personal computers on the shop floor and linked them to the laboratory where metal samples are analyzed. Feedback via the computer network enables hourly workers to customize batches by putting into each ladle just the right amounts of additives. This would be impossible without the help of the computer system, according to Leach. For example, to make magnesium ferrosilicon products, as many as 11 different additives and 5 sieve distributions must be controlled to tailor materials to particular specs.

The new system allowed Globe to match customers' requirements with far greater precision. Scores for a key statistical measure (CPk) had frequently been under 1.0, which made the material unacceptable to some buyers. Now typical values are over 2.0, and sometimes they range as high as 10.0. Even demanding customers began to drop requirements for 100% testing. Globe spent $60,000 on the computer system but because of the reduced need for inspections now saves some $600,000 annually, Leach estimates. In addition, now that customers get highly customized, high quality products, competition finds it much tougher to encroach on Globe's markets.

Getting a highly motivated, committed work force was another problem Globe faced. In 1985, as new management started the shift to a total quality program, distrust by the union at the Beverly plant escalated into a strike in 1986. The union was eventually decertified. Many workers were called back to work but without the restrictive work rules that had impeded past efforts to improve systems, according to Leach.

The union at the Selma plant, however, cooperated enthusiastically in the shift to a new quality system.

Management at the Selma plant made various changes to improve employee morale, such as instituting benefit and pension plans for hourly workers matching those of salaried employees, eliminating time clocks, and establishing a new profit-sharing plan that could provide bonuses of up to $5,000 a year. Still, a major effort was needed to eliminate the former autocratic management style. To make the shift to continuous improvement led by workers rather than management, Globe set up four types of team-based quality. The total quality effort is led by a Quality–Efficiency–Cost (QEC) steering committee which includes a full-time coordinator for quality circles. Following are the four types of teams:

1. *Departmental teams.* These teams meet for an hour each week, before or after a shift. Members are paid an hour overtime for the meetings, which are chaired by an hourly worker. The team leader is trained to maintain each meeting's focus on generating ideas for improvements in quality, efficiency, and costs. Preprinted forms help the seven-member teams to generate and discuss about ten ideas per meeting. The ideas go to the coordinator for consideration by the QEC committee, which decides whether the idea will be implemented right away or studied further. The plant manager has the authority to act immediately on a good idea. Sometimes workers see changes made as a result of their suggestions by the shift following a departmental team meeting.

2. *Interdepartmental teams.* These teams also involve hourly workers but from various departments. Their focus is on improving interaction and solving problems among groups handling different parts of the manufacturing process.

3. *Project teams.* These teams are set up as needed to deal with particular phases of Globe's operations. In the past, only slaaried workers carried out such efforts, but now hourly workers are involved as well. Taguchi methods for designing manufacturing experiments, storyboarding brainstorming methods, and advanced statistical measures are used. Formerly, managers initiated changes in work methods and dictated them to each department. Workers often resisted. Now participation makes workers much more cooperative in implementing new approaches.

4. *Interplant teams.* These teams, which meet once a month, include hourly workers from the Selma and Beverly plants. In the past, workers from one plant were not even permitted to go through the other production facility. The work of these teams quickly resulted in a wide range of technology exchanges between the two plants. Workers serving on the teams rotate in order to expand the exchange of ideas.

In developing quality plans, the teams use a variety of techniques, including the familiar Pareto charts (bar charts in which failure modes are ranked so the most severe can be tackled first), fishbone diagrams (for aid in tracking defects back to root causes), and statistical techniques. They also use imaginative storyboarding methods

～ WHAT SHOULD GO INTO A QUALITY MANUAL?

A key element of Globe Metallurgical's world-class status is a quality manual that provides consistency throughout its operations. Here's what the manual includes:

- *Procedures.* These are instructions for workers in each function involved in an operation involving a set of tasks. The procedure for sampling molten magnesium ferrosilicon, for example, may include separate instructions for the crane operator who pours the molten metal, the furnace tapper who oversees ladle operations, and the tapper's helper.
- *Job work instructions.* While the procedures give a comprehensive view of a total operation, each participant is referred to specific job work instructions, which are much more detailed explanations of how specific tasks should be performed. This ensures consistency across different shifts and plants.
- *Critical process variables.* These variables are identified for each control point within a process. Process capability can be assessed as variability is reduced, and generally the result is expressed in terms of the CPk statistic. Critical process variables

(CPVs) can be added or deleted from the control process by the QEC committee.

- *Product parameters.* These are developed to meet each customers' quality requirements or specifications as developed in a control plan agreed on by Globe and the buyer. The plan not only defines output targets but identifies the CPVs that Globe will monitor to achieve control. Most customers now require control charts to accompany shipments.
- *Failure-mode effects analysis.* Every fundamental aspect of an operation is analyzed to determine the risk of failure. Scores of 1 (lowest risk) to 10 (highest risk) are assigned to the occurrence rate, detection capability, and severity to the customer for each factor. The three risk scores for each factor are multiplied together to obtain a risk priority number (RPN), which can range up to 1,000 for a factor with maximum risk ($10 \times 10 \times 10$). Once factors with high risks of failure have been identified, the company can set up process and engineered controls to reduce the potential for defects.

borrowed from Walt Disney to track through the actions needed to control key process variables.

The team concept has also helped Globe solve a problem with its materials suppliers. Although the company provided training to vendors and described how they should apply SPC, some were slow to respond. So Globe sent QEC teams to supplier locations to help workers there to implement total quality approaches. This effort resulted in a huge payoff. One example illustrates just one of the many important gains resulting from the supplier program. For many years, a supplier of quartzite had allowed small pellets, undesirable for silicon metal production, to be mixed into railcar shipments. Despite complaints from Globe management, the problem persisted, but the supplier was retained because there are few sources for low-impurity quartzite. Instead, the smaller pieces were screened out and used in another process where much lower-cost material would have sufficed. To solve the problem, a QEC team from Globe visited the supplier's site to help workers there learn statistical methods, and

SPC charts were then required with all shipments. As a result, undersized pieces of quartzite were completely eliminated from this supplier's shipments. It cost Globe about $2,500 to send the QEC team, and it is estimated that savings as a result of these efforts are some $250,000 a year, according to Leach.

Globe now has its own certification program and audits its suppliers once a year. Certification leads to longer-term contracts and enables the development of a partnership relationship. As process capabilities are proven, Globe may waive requirements for control chart reports for each shipment. Also, consistent quality from suppliers has enabled Globe to step up just-in-time (JIT) deliveries of materials to its plants, cutting inventory costs.

Aside from Ford, other Globe customers also began to require audits of Globe's quality systems, sometimes with different requirements. Rather than bridling at these multipronged demands on a small firm, the company broadened its total quality approach to satisfy any such requests. At one time, General Motors' foundry, for example, set up a JIT system, but since GM did not want to risk a shutdown in case of defective material or a missed delivery, it required Globe to maintain a ten-shipment inventory in a nearby warehouse. General Motors sent its own personnel to check quality even though Globe provided 100% checking. By demonstrating to GM that its quality systems ensured dependability, Globe built enough trust that the warehousing requirement was eliminated.

This small firm won much more than a Baldrige Award for its strenuous efforts. All over the world today in silicon metal and ferrosilicon markets, customers require "Globe-quality" silicon and ferrosilicon products. That is world-class!

～ Wallace: Tapping America's Hidden Asset—the Work Force

In the mid-1980s, Wallace Co., Inc., a Houston, Texas, distributor of pipes, valves, fittings, and other products for the chemical and petrochemical industries, faced a bleak future. New construction in the industries it served was in the doldrums, with little prospect for a pickup, and this was the main source of its business. Adding to the grim outlook was the assignment of Wallace to a high-risk group for safety due to excessive violations of OSHA regulations.

Facing what appeared to be imminent oblivion in 1985, the small family-owned company launched a continuous quality improvement program to help turn things around. As a result of the quality push, sales nearly doubled over the period from 1987 to 1990, growing from $52 million to $90 million, and market share expanded from 10.4% to 18%. In 1990, as a result of its success in transforming the company using total quality principles, Wallace was awarded a Malcolm Baldrige National Quality Award.

How did Wallace manage not just to survive but actually to thrive in spite of a market that remained stagnant throughout the late 80s? The most important contribution, according to Michael E. Spiess, executive vice president and chief operating officer, came from a hidden asset that is underutilized in most U.S. companies—the work force. As management began to solicit ideas from workers at all levels, it found wide opportunities to make use of the company's skills for a major shift toward supplying maintenance and service operations while greatly lessening dependence on the static market for new construction. Wallace was not only able to deliver replacement pipe, valving, and pumps as needed, it also began to do installations which could be guaranteed because of the expertise and dedication to quality of its work crews.

Getting this transfusion of business-saving ideas had to start from the top. As the company embarked on its total quality journey, John W. Wallace, chief executive officer, gave his pledge to the work force that, although every opportunity for stream-lining operations needed to be explored, no one would lose her or his job as a result of the program. The five top officers of the company then underwent more than 200 hours each of intensive training in the principles of continuous improvement, including statistical process control.

Wallace initiated benchmarking of other firms with outstanding total quality systems and decided to adopt Monsanto's program as a model. A task force was formed to direct the total quality effort, including six customers, six manufacturers, and six Wallace product managers. The company developed a quality mission statement with input from all 280 of its employees, now called associates.

The entire culture of the company changed from reactive correction to proactive prevention. To define directions for total quality initiatives, the company developed 16 quality strategic objectives, several of them centered on improving customer satisfaction, and organized voluntary teams to come up with plans for achieving improvement in each area. The initial 16 strategic objectives were

- leadership development
- quality business plan
- on-the-job training
- information analysis
- statistical process control
- quality education
- human resource development
- quality improvement process/team involvement
- customer service/satisfaction
- employee reinforcement/incentive
- *Quality Pathfinder* (a newsletter)
- suggestion system
- internal audit
- internal/external benchmarking
- vendor quality improvement plan
- community outreach

The voluntary team concept evolved from quality circles in 1985 to a wide array of fully empowered quality improvement process teams. These teams take on assignments to solve SPC problems, streamline job processes, monitor cycle-time reduction, study failure mode and effects analysis, and help all divisions improve processes. Each team is headed by an SPC coordinator, who must pass a test to certify skills after having a minimum of 274 hours of SPC and team training. All Wallace associates now participate in at least one of the teams, which normally meet every 1 or 2 weeks during working hours. If after-hours meetings are held, participants are paid overtime.

∼ Wallace's Quality Mission Statement Says It All

One way to ensure that quality takes the spotlight in all corporate activities is by means of a clear, uncompromising mission statement. Following is the mission statement developed by Baldrige Award winner Wallace Company with input from all its employees (called associates). The statement is issued to all associates, customers, and suppliers.

We, the Wallace Co., Inc., are ethically and operationally guided by the Principles of Quality outlined within this document. With these principles, we have built our company; with these principles, we offer leadership for tomorrow . . . Our commitment is to provide Quality products at fair prices with the highest level of service, honesty, and integrity for our customers.

. . . We are committed to Continuous Quality Improvement, Customer-Driven Service, Product Quality, Statistical Process Control, and Employee Involvement.

Commit to Innovation

. . . We want people to feel pride in their work; we are committed to recognizing the contributions and service of our employees. We are an employee-oriented company; we seek two-way communication in daily interactions; we drive fear from the workplace.

Commit to Training for All Employees

. . . We pledge our support to enable all employees to: Understand and meet the needs of our customers, both internal and external. Utilize the tools of Statistical Process Control. Work for Continuous Improvement of all systems and processes.

Commit to Leadership

. . . We know that "Leadership" means working closely with all employees, endorsing pride of workmanship, listening to and responding to the suggestions of those closest to the job, building in Quality in all aspects of operations, focusing performance on the positive, and developing a team spirit. We are in a Partnership of Excellence in our business relationships; we accept tomorrow's challenge.

Commit to Long Term

. . . We work with both suppliers and customers toward the goal of assuring Quality products and Quality service. We realize that Excellence of Product and Service is achieved through the ongoing teamwork between Wallace Co., Inc., its custoemrs, and its suppliers.

The key to this teamwork is communication. We listen to our employees; we listen to our customers; we listen to our suppliers. We respond to their needs.

Commit to Continuous Improvement

. . . We know that improving procedures requires a team effort; we combine the expertise of people closest to the job with knowledge of the people who design and review procedures. In all company procedures, we "Build Quality In" and pledge company resources to help our people "Do it right the first time."

Commit to the Future

. . . We focus upon the needs of our customers. We adhere to Quality as our standard. To this end, we publish this Quality Mission Statement for distribution to all. We live the Wallace way of life—a business life consistent with our values.

One of the teams is a safety quality assurance team with the goal not just of eliminating excessive OSHA violations but of setting a new safety standard in the industry. Wallace's quality safety alert program, developed with the help of this team, received high praise from the company's insurance carrier, and the program has been called "leading edge" by the Texas Safety Association.

Wallace recognizes that simply empowering workers falls far short of what is needed to achieve a cultural revolution. Clear communication is necessary to convince workers of top management's sincerity and commitment to quality. They also need to understand corporate strategies to help them in setting tough goals that will optimize their efforts. They need extensive training not just in their assigned tasks but in the concepts and principles of total quality.

In the 1988-1990 period, Wallace spent more than $750,000 on training, which included such subjects as:

- quality awareness
- statistical process control and SPC problem-solving
- process capabilities and failure mode and effects analysis
- data collection
- sales and customer service
- leadership, coaching, and communication/motivation
- product knowledge and cycle-time reduction
- benchmarking and vendor certification
- quality business planning and information analysis

The training is targeted to appropriate groups. All associates, for example, learn about quality awareness, but only those who must analyze processes take detailed SPC training.

Through weekly teleconferences, the accomplishments of various teams in a "quality wins" program are cited and given credit for suggestions. John Wallace, the CEO, sends letters of congratulations to associates' homes when appropriate. Dinners, picnics, and similar events are awarded to deserving teams. Benchmarking of other quality programs, including Globe Metallurgical, continues at Wallace, and as a result of studying Milliken's recognition program, Wallace has started to hold "sharing days" to help build teamwork and participation.

One result of these and other special benefits programs for associates was that from 1987 to 1989 absenteeism fell from 1.4% to 0.7%, less than half the industry average, and turnover fell from about 7% to less than 3%. Sales per employee have risen more than 66% since the program started and are now the highest in the industry. Inventory turns are now 4.3, which is also close to the highest in the industry, according to chief operating officer Spiess.

~ Improving Customer Satisfaction Gets In-Depth Treatment

Pivotal to Wallace's growth in market share has been its dedication to better serving customers. Customer surveys analyzed by Pareto analysis (using bar charts to rank factors in descending order) revealed that the three most important measures of performance were

1. on-time delivery performance
2. complete and accurate shipments
3. error-free transactions.

Wallace was able to identify 72 discrete processes that contributed to the most important of these, on-time delivery. They ranged from answering the telephone promptly to maintaining optimum inventories. Each process was then broken into steps and subjected to Pareto analysis. The most critical steps in achieving on-time delivery were identified as properly completing a sales order, filling the order, receiving materials, maintaining optimum inventories, locating materials, delivering the correct material, and maintaining accurate customer files. Just these seven factors contributed 85% of total process variation, so they were tackled first. Implementing them involved 81% of all associates.

To determine progress, the SPC coordinators maintain data concerning each of these factors at each district office and on the firm's mainframe computer. In-depth study of each factor determined that the definition of on-time delivery differed among various customers, so Wallace developed customer profiling. A wide variety of profile information unique to each account is now on line and available to anyone in the company. On-time deliveries moved from 75% to over 92%, and major accounts receive written guarantees of at least 95% on-time deliveries.

Work force participation has led to a delivery concept that sets the company apart from its competitors. Wallace truck drivers decided they could improve on industry standards, which call for simply delivering a shipment to a receiving dock. Instead, if no union or work rules are violated, the Wallace truck driver will make deliveries to specific locations within a plant or plant complex and then stay to help unload the shipment, ensure that it is properly stowed, and update inventory counts.

"The main objective of total quality management is total customer satisfaction," according to Paul Vita, Wallace's director of national accounts. Keeping in touch with customer's changing needs, however, means doing a lot of listening and providing those who must interface with customers quick access to the information they need to provide accurate answers and prompt service.

Even that isn't enough, according to Vita. A supplier must anticipate change and adapt to new needs. It is essential to keep coming up with new services and product offerings to plumb the changing requirements of the marketplace. One way that Wallace finds new supply and service opportunities is by studying changes over time revealed by the detailed customer profiles. Although Wallace has been constantly tracking customer profiles over the past 15 years, Vita points out that a major shift since 1986 from supplying engineering construction firms to supplying service and maintenance operations has changed the customer base dramatically. Trying many new options guides Wallace in reshaping its offerings to better serve existing buyers as well as in determining the needs of new customer groups. "Throwing out multiple new ideas continuously also helps to confuse the competition," Vita adds.

Like many companies, Wallace takes a variety of surveys to find out how customers rate it versus its competitors. In addition, Wallace performs studies aimed at discovering specific steps it might take to improve. One study, for example, asked customers to rate Wallace's performance on a scale of 1 (poor) to 5 (the best) for each process involved in its business. In the latest study of this kind scores averaged 4.3.

A vital aspect of customer relations is the ability to respond quickly to any complaints with accurate and complete information. Wallace guarantees customers a response to their concerns or complaints within an hour. To do this, it initiated a total customer response network and trained associates to resolve problems through a streamlined process. Every effort is made to turn the few negatives that do occur into positives through quick, fair resolution of any difficulty. Any complaint also initiates a process of root cause analysis, so appropriate recommendations reach a quality management steering committee. This feedback helps to determine areas for improvement in customer service to prevent further errors.

Critical to such efforts is an effective technological infrastructure. Workers have on-line computer access from anywhere in the organization to obtain information about inventories, customer profiles, prices, and so on. The computerized information system installed in 1964 has been upgraded periodically, most recently in 1988-89. Speed of response comes from using on-line data rather than paper whenever possible. The alphanumerics for part numbers, for example, now all have meaning. By using generic part numbers, sales offices find it much easier to make quick, accurate price quotes, and the standardized coding systems cut down on errors. Information on availability and properties of unique products, such as specialized pipe, is also instantly available on line.

Keeping tabs on operations via the computer network provides a data base that supplies information on service performance trends, for statistical process control charts and analysis handled by SPC Coordinators at each district office. The validity of the data is verified through periodic checks with customers, a perpetual inventory management system, and an annual financial audit.

Collected data can also help in responding to customer requests. If a customer wants to see material test reports on carbon steel pipe that was sold 6 months earlier, for example, the data can be accessed, printed out, and immediately faxed to the customer's location.

To make it as convenient as possible for customers to do business with Wallace, electronic hook-ups are being extended to bring them right into the information loop. This includes fax machines, car phones, special toll-free numbers, and special conference telephone connections. The company also wants more of its computer resources to be available to customers. To accomplish this, Wallace not only has tried to maximize the use of electronic data interchange (EDI), it has also assisted many customers in implementing and piloting their own EDI programs. Now more than 40% of incoming orders are handled via EDI, according to John Wallace. The on-line real-time inventory management systems enable buyers to get accurate, up-to-the-minute quotes and expediting information on their orders.

Along with the effective use of technology, Wallace has worked to improve its personal relationships with customers. Every associate, for example, is required to spend at least one day a year at a customer's site, gaining first-hand knowledge of that firm's service needs. For major accounts, Wallace uses an approach called team approach selling, in which it assigns to each account a team that includes outside and inside sales people, a sales assistant, an order filler, and an accounts receivable person. Every associate is authorized to make decisions on the spot costing up to $1,000, and may even make higher-value decisions if senior management is not immediately available.

To ensure that it continues to improve its service to customers, Wallace also had to ensure that its suppliers were doing an outstanding job. With guidance from Monsanto, the company set up its own vendor certification program, and supplier quality is constantly monitored. Wallace reduced its supplier base from more than 2,000 to 325 over a 4 year period and developed partnerships with key vendors. the company provides suppliers with week-long total quality management training courses to help manufacturers improve their own quality. It also serves as an intermediary when customers go to new technology and need to link it with the firms making the products Wallace supplies. On-time deliveries to Wallace from its vendors is now close to 90%.

The credit for Wallace's turnaround must go to its workers, says chief operating officer Spiess.

"If you free, nurture, and listen to people, they will lead you in the right direction," he says, adding that "one of the saddest things in American industry is the underutilization of the work force." Because Wallace's ownership allowed its people to stretch themselves, he says, "we achieved what was thought to be impossible."

In coordinating the total quality effort at Wallace, Spiess learned what he calls "the three truths of quality, which is much more than a word, but an all-encompassing way of doing business that leads to total customer satisfaction."

Following are Spiess's three truths:

1. Quality is not magic, it is common sense.
2. Quality is good for business, because it eliminates waste and rework and energizes the work force.
3. Quality requires full commitment from everyone in the organization.

∽ Marlow Industries: Convincing Other Managers Proves Critical

Adding encouragement for hundreds of small firms hoping to someday win a coveted Baldrige Award was the 1991 victory of Marlow Industries, Inc., a tiny, $13 million-a-year maker of thermoelectric cooling devices with only 155 employees.

Just getting started proved to be one of the hardest things this privately owned Dallas-based company went through in moving toward total quality. When Ray Marlow, a former Texas Instruments engineer, founded the company in 1978 he wanted quality to be its cornerstone. At that time in the United States *quality* was viewed in the narrow sense of product quality, and achieving it was traditionally delegated to a specialized department. As Marlow recognized the much broader concept involved in *total quality*, however, he realized it would have to become part of everything the company did. Having come from a large, global company, he saw that smaller suppliers would also have to become world class to help the larger companies they sold to as well as to ensure their own products and services would remain competitive.

The total quality process started in 1986, after a 2-day meeting with a major customer, Hughes Aircraft, in which Marlow's managers learned about statistical process control (SPC). Seeing the need to get wide participation, Ray Marlow began to work directly with factory workers. He tried to instill in them the spirit of continuous improvement based on SPC.

It didn't work very well. After some 5 months of preaching the total quality gospel, Marlow began to get complaints from middle managers who felt left out of the process. He then went on a campaign to convince all his managers to join the quest. It took another year before even one of them came around.

That was the beginning of a complete transformation. By 1991, all of Marlow Industries' executives were so much a part of the total quality drive that each of them personally filled out a section of the winning application. Without the resources of larger firms, these managers had to do much of the paperwork after hours and on weekends. Being a small company did have one major advantage, however. Managers who listed a weakness in their section of the application could then "just go and fix it," says chief operating officer Chris Whitzke.

Constant communication with employees and support for team efforts by upper management are key elements in Marlow Industries' success. "No one is afraid to speak up when we walk into their areas," says Ray Marlow. Each new employee, even a temporary, is greeted within the first hour by a senior executive and encouraged to exchange ideas openly.

Teams at Marlow Industries are given more autonomy than in some team-based operations. If a quality goal is achieved, for example, the team has the freedom to go out for a pizza celebration without management permission for the expense. Cost is an excuse that some small companies use in avoiding a transformation to a total quality culture; but, according to Ray Marlow, it takes more time than money to make the total quality approach work. Great progress can be achieved without excessive spending, he feels. At Marlow, for example, internal quality training was budgeted at some $120,000 in 1991, with another $20,000 for suppliers, adding up to only about 1% of total annual revenues.

Marlow himself has become such a convert to the total quality approach that he has campaigned for formal quality programs at the American Electronics Association. When he had trouble with his American luxury car, Marlow convinced his local automobile dealer to institute a total quality approach at his dealership.

According to chief operating officer Whitzke, it's that kind of personal commitment by the CEO that's needed to make a small firm a strong candidate for a Baldrige Award.

Chapter 8

Federal Express Wins in the Tough Services Category

Achieving total quality is a challenge for manufacturers, but it's even tougher for service organizations. A manufacturing company can carefully monitor its processes. It can chart measurements at each step to provide the feedback needed to gain control of production by steadily reducing variation. Once it has achieved stability, it can more easily identify the root causes for any defects, rank them in order of importance, and correct them. It can continuously improve product quality by gradually narrowing the range of allowable process variation and repeatedly eliminating any remaining causes of trouble.

Services, however, are intangibles. The processes required to achieve excellence are harder to monitor and, thus, not as readily controllable. Even manufacturers often have a much harder time getting a handle on the services side of their business than they do in managing factory operations. Executives may be unaware when trouble is brewing because they can't look over an employee's shoulder every time a transaction occurs. Disgruntled customers may know what's wrong, but instead of complaining, most of them will simply take their business elsewhere. As a result, a sharp drop in market share may be the first sign of eroding service.

Finding reasons for the trouble can be a maddening exercise. It's hard to measure each step in service delivery, making it tough to reduce variation. Aside from occasional complaints, there are no defective products for statistical analysis. That makes it much more difficult to identify and rank root causes. Customer surveys help, but their reliability falls far short of that provided by careful analysis of defective hardware. Lacking good feedback, how can a service company judge the effectiveness of attempted improvements?

Without clear proof of steadily improving service levels, a company has no hope of winning a Baldrige award. The scoring depends too heavily on the demonstrated success of continuous improvement efforts over an extended period. The difficulties service companies have in clearly demonstrating continuous improvement may be one reason it took 3 years of the Baldrige competition, with 6 winners in the large manufacturer category (the maximum number allowed by the rules), before even a single service organization could make the grade. Finally, in 1990 Federal Express Corporation broke the ice.

The Memphis, Tennessee–based package delivery company, famed for its fleet of trucks and aircraft with their distinctive purple markings, became the first National Baldrige Award winner in the service category.

Because meeting the Baldrige criteria is so difficult for a service company, it would seem that only a highly exceptional total quality program could make the grade. And Federal Express's quality program is exceptional indeed. Formidable obstacles face any service company trying to achieve total quality, whether the firm is in transportation, retailing, distribution, publishing, entertainment, insurance, or any other business dealing in intangibles, but Federal Express had an additional burden. As a pioneer in overnight parcel delivery, it had to invent the business every step of the way. When the company was founded in 1973, the concept of a private fleet of planes and trucks successfully moving items overnight all over the United States through a hub-and-spoke network seemed highly improbable. Making a profit at it seemed almost unimaginable.

Yet somehow, through very tough times, Federal Express not only succeeded, it thrived. Recently it has even expanded its concept to cover the globe. It has grown to a $7 billion company with 94,000 employees, more than 400 aircraft, and 30,000 pickup and delivery vans handling some 1.5 million packages a day. Maintaining a broad view of quality, for both internal and external customers, played a big part in the Federal Express miracle. Many thousands of service organizations in the United States can learn vital lessons from the accomplishments of this pathfinding company.

～ Federal Express: Recognizing the Importance of Information

When Federal Express Corporation was founded in Memphis, Tennessee, in 1973, Frederick W. Smith, chairman and CEO, decided that even though it would be a remarkable achievement to set up a national overnight package delivery service in the United States, it wouldn't be enough. Smith recognized that the company would also have to be in the information business.

The rapid rise of fast-paced high technology businesses created optimism for the overnight delivery concept. High tech companies needed to move swiftly to capitalize on market opportunities, so time would be critical in shipping complex drawings, plans, and vital components. Organizations dependent on high tech equipment also would need rapid delivery of spare parts in case of any breakdowns. To attract such business, Smith recognized that the service would have to build a strong reputation for dependability. Achieving that reputation would require highly effective methods for finding and correcting any hitches in picking up, transporting, sorting, and delivering documents and packages. That would require monitoring of far-flung operations and careful analysis of results to find areas that needed improvement and to ensure that fixes were working as intended.

It became clear that data useful for internal operations were not sufficient. What if someone called in to check on the status of a vital shipment? Was it adequate to have a clerk check a schedule that said that Route X's load should have gone out on the 9 P.M. flight to Memphis? That wasn't very reassuring. Smith realized that Federal Express has what he called a "custodial" responsibility. It would be vital to track *each* package and to provide customer service representatives with immediate access to the latest information. Somehow there had to be positive feedback, indicating just where any particular package was within the system at all times. Certainly such information would be useful in monitoring and improving the company's efficiency. Even more important, however, each customer who called about a specific package would be reassured by an immediate, responsible, informed answer to any query.

The concept sounded simple, but implementing it proved to be a difficult, costly process that still continues. To develop the unique electronic capabilities needed to gather data on each package as it moves through the network, Federal Express eventually established its own research and development laboratory in Colorado Springs, which now employs more than 400 people. Processing and distributing the needed information has made the company one of IBM's largest customers.

Gathering such detailed intelligence on all its operations gives Federal Express an important edge over most service companies. It can monitor each process step, from aircraft maintenance procedures to the performance of its vans in pick ups or deliveries. Individual breakdowns in the system can be analyzed so that continuous improvements can be made, and the results of improvement efforts can be tracked.

The power of the system is illustrated by a recent incident in which a customer service representative called from L.L. Bean, the Freeport, Maine, catalog-marketer of outdoor gear. A Bean customer was on the line complaining that a delivery had not arrived even though it had been shipped the day before via Federal Express. After looking at her computer screen, the FedEx representative said: "Ask your customer if it's been raining there."

"Yes, it has," came back the reply.

"Then have her look under the back porch." The package was found just where a message from the delivery van to Federal Express's COSMOS (customer service master on-line system) data base indicated it would be.

Such an information-rich environment enabled Federal Express to provide concrete proof of steady improvements in its delivery of services. Without supporting data, the company probably could not have won a Baldrige Award in spite of the excellence of its total quality approach. Still, no matter how powerful the computers, they can only tabulate and report outstanding results. Federal Express has been successful at achieving ever tougher goals because of an outstanding, innovative approach to total quality that actively involves everyone in the organization.

~ Quality: "Doing Right Things Right"

The direction for the program starts with the company's definition of quality: not just "doing things right." but "doing the *right* things right the first time . . . every time." The message goes beyond emphasizing that each job must always be done well. The work force learns that it is also vital to consider whether *what* one does can be improved. This sets the stage for contributions to continuous improvement efforts from everyone in the organization.

The corporate philosophy provides a focus for improvements. The company's mission statement is more elaborate than those of many firms, but it details the elements considered essential for world-class performance.

Federal Express is committed to our People–Service–Profit philosophy. We will produce outstanding financial returns by providing totally reliable, competitively superior global air–ground transportation of high priority goods and documents that require rapid, time-certain delivery. Equally important, positive control of each package will be maintained, utilizing real time electronic tracking and tracing systems. A complete record of each shipment and delivery will be presented with our request for payment. We will be helpful, courteous and professional to each other and the public. We will strive to have a satisfied customer at the end of each transaction.

In communicating to employees, Federal Express frequently emphasizes answers to what is assumed to be everyone's silent question: "What's in it for me?" The result of doing the right things right the first time, it is explained, will be that the job becomes easier and the quality of work life will be improved because there will be fewer complaints from customers and less hassle and rework. Workers are assured that they will have an opportunity to participate in problem solving and to win team awards.

Implicit in the people–service–profit (PSP) philosophy, the work force learns, is that the effects on the firm's people are always considered first when decisions are made. Management's rationale is that taking good care of employees will lead them to deliver superior service to customers, thus encouraging further use of the company's offerings. This, in turn, will provide the profits that will allow the company and its employees to thrive.

Adding even more specificity to the mission, a 100% performance target is clearly stated.

We are committed to delivering each shipment entrusted to us on schedule 100% of the time. Equally important, we will maintain 100% accuracy of all information pertaining to each item we carry. Our objective is to have a 100% satisfied customer at the end of each transaction."

Many firms issue credos and post slogans on the walls. But Federal Express ensures that employees are aware of progress toward the tough goals laid out in company statements. Service quality indicators (SQI) give numerical measures of performance, and an SQI index combining scores for 12 indicators is constructed, reported weekly, and summarized monthly to show progress toward the 100% targets. Scores for the 12 indicators currently used to calculate the SQI index are obtained by summing a weighted average of the daily failure points for each component (see the box "Service Quality Indicators"). The weighting factors, ranging from 1 to 10, are chosen to emphasize the customer's view of service. Every time a phone call comes in to one of the 16 domestic call centers, for example, the goal is to answer it within four rings. Any call reaching six rings is counted as a service failure.

Although providing summary information helps give employees a sense of the progress of the total organization, it doesn't provide much insight into where actual trouble is occurring and what is being done to improve the system. In some compa-

~ SERVICE QUALITY INDICATORS: HOW FEDEX MEASURES ITS PERFORMANCE

Customers are the best judges of the quality of services. That's why, in developing a composite quality indicator, Federal Express looked for factors reflecting its customers' view of performance. It identified twelve components as key elements in successfully delivering what customers want. Some failures have much more impact than others. Losing or damaging a package, for example, is much more serious than simply delivering one a few minutes late. Therefore, the company assigns weighting factors according to the customer's perception of their importance.

The service quality indicator (SQI) is the weighted sum of the average daily failure points for the 12 components, and it is reported weekly and summarized monthly. Some 60 million weighted opportunities for error exist each day, yet SQI scores have steadily dropped until they now run about 0.4 of 1%. The company now calculates a similar SQI figure for the international delivery service. The purpose of the service quality indicator is to help Federal Express identify and eliminate causes of failures but *not* to place blame. If courier mislabeling of packages was found to be a cause for wrong-day-late failures, for example, the SQI team would work on creating effective methods for preventing miscoding at the source rather than on developing an elaborate and expensive expediting system. Finding out what dissatisfies customers is a first step, but then a cooperative effort within the context of all the goals of the organization is needed to find optimum solutions to any problems—so that employees do the *right* things right.

Following are the SQI components:

Failure	Weight	Description
Right day late	1	Delivery after the commitment time but the right day.
Wrong day late	5	Delivered the wrong day.
Traces unanswered by COSMOS	1	Number of proof of performance requests by customers where exception information, proper scans, or proof of delivery (POD) data are not in COSMOS.
Complaints reopened by customers	5	Includes customer complaints on traces, invoice adjustments, missed pickups, etc., reopened.
Missing proof of performance	1	Billing documents that don't match a POD in COSMOS or from the field POD queue on a timely basis, including prepaid and metered packages as well as those that are invoiced.
Invoice adjustment requested	1	Packages on which customers request an invoice adjustment, including those not granted because a request indicates the customer perceives a problem.
Missed pick-ups	10	Complaints from customers recorded as missed pickups.
Damaged packages	10	Includes packages with either visible or concealed damage and weather and water damage. Also includes contents spoiled or damaged due to a missed pickup or late delivery.
Lost packages	10	Includes both packages missing and those with contents missing due to pilferage.
Overgoods	5	Packages received in Lost & Found (no label or identifying data inside package).
Abandoned calls	5	Any call to FedEx that is not answered, which is any call in which caller hangs up without speaking to an agent after 20 seconds from receipt of call.
International	1	Includes components from the performance measurement of international operations.

nies, access to this sort of negative information is restricted to analysts and managers. But not at Federal Express. Using its own television network, FXTV, Federal Express chronicles any gliches in the system, such as delays or sorting problems, on video every work night and broadcasts them all over the organization via satellite for taping at 4:30 A.M. If the report shows, for example, that the Alaskan flight arrived 12 minutes late, the audio explains reasons for the failure. As workers come into the superhub in Memphis, where about a million packages each night are transferred from incoming to outgoing flights, they pass a row of TV monitors displaying details on any foul-ups in the previous night's performance. The taped video can also be shown when convenient at other facilities throughout the system—more than 1,600 of them worldwide.

Any FedEx facility can see the overall presentation, but some groups may wish to call up additional details of special concern, perhaps plans for improving aircraft maintenance procedures. Each facility holds a service meeting at 8:30 each morning, where possible solutions to any system problems of the night before are presented. Some people work through the night to ensure that potential fixes will be ready for presentation by the morning meeting.

Package volume has grown to more than 1.5 million parcels a day, generating data that is beginning to overload the company's computers. Already Federal Express has become one of IBM's biggest customers, with one of the largest information data bases in the world. Rather than cutting back on information requirements, however, Federal Express plans a major expansion of its computer facilities while adding to both the kinds and amount of data to be collected. The company was one of the first sites to receive IBM's giant new Enterprise systems.

Federal Express's own R&D labs in Colorado Springs are constatly developing unique electronic systems, and the company does its own information systems R&D at data centers in Memphis and Los Angeles, the headquarters of Flying Tigers International, which Federal Express acquired in 1989.

∽ Supertracker: Following Packages from Pickup to Delivery

No area better illustrates Federal Express's dogged commitment to self-improvement than the effort to track each package through the system and to ensure that the data are immediately accessible to respond to customer queries. In the early days, couriers had to feed back information and get pickup assignments from pay phones along their routes. But in the late 1970s swelling package volume caused too many calls to flood switchboards in some cities. The company began to seek ways to use technology to link the couriers to the information system loop right from their vans. Bar-code technology appeared to be a promising candidate.

The initial goal was to develop a hand-held terminal (HHT) that could scan an identification number for each package starting right at pickup and then following it through each transfer point to final delivery. Couriers would use a small keyboard to enter data on destination, sender, type of service, and so on. This information would be transmitted to the COSMOS data base where it could be immediately accessed using an individual package's identity number. The first commercial devices adapted for this purpose, using software developed by Federal Express and programmed onto a fixed memory chip, were bulky and cumbersome and proved not to be rugged enough for use in vans. This meant that packages could be scanned only at transfer stations but not in the vans.

Rather than waiting for a more rugged, compact hand-held device to be developed, the program was split into two parts, dubbed COSMOS IIA for the station-only scanners and COSMOS IIB for a future smaller, ruggedized version that could be used in the field by couriers. This evolutionary approach, according to Chris Demos, business advisor–integrated systems, follows a company philosophy frequently invoked in development projects: "Don't let *best* get in the way of *better*."

At first things did not go well. The scanning process required extra personnel at each station and slowed down the transfer of packages. Some transfer stations did not scan as diligently as others, so compliance reports had to be generated to jog the station managers into ensuring that all packages were scanned. Adapting the system to any special requirements or changes in the system was a cumbersome process. Federal Express had to go back to the terminal supplier to revise the software and then to program new insertable memory chips (called PROMs, for programmable read-only memory). It was difficult to distribute the revised chips and to ensure that every scanner had an up-to-date PROM installed.

By mid-1982, the burgeoning volume of pickup calls in some cities, especially at the end of the day, led to a shift to microwave transmission of data from the local station's computer to the vans using a new system called digitally assisted dispatch system (DADS). Displays on the DADS terminals showed the couriers where pickups had to be made. This intensified the need for a lightweight scanning device that could be carried in the vans and used to link information at pickup and delivery points right into the COSMOS data base. Such a system would at last provide what CEO Smith had envisioned as a custodial system for positively tracking each package from pickup to delivery and making up-to-date data available for concerned customers.

Federal Express designers worked with a small outside firm and various testing laboratories to develop the unique technology required. After extensive testing, including active use by couriers in the field and numerous redesigns, a device called a Tracker went into use in 1985. Video presentations, including interviews with couriers who tested the Tracker, helped gain acceptance for its use throughout the system. Still, since it did not offer any special advantage for pickups, couriers only used hand-held Trackers to record deliveries.

The hand-held scanner fit into a shirt pocket, could withstand extreme temperatures and heavy rain, and could be dropped from 7 feet without damage, all major goals of the design project. But engineers continued to improve the Tracker. In early versions, 12 queries presented in a small display window had to be answered if there was any hitch in making a delivery (such as a business being closed). Later the software was modified so that once the type of exception was specified, the courier would need to answer as few as three queries, cutting down on the time needed for data entry.

Meanwhile new regional hubs were added, including Newark, Indianapolis, and Oakland, for aircraft transfers in the eastern, midwestern, and western regions, respectively, and a number of smaller trucking hubs for shorter range deliveries. But each station continued to route all its packages the same way. With suitable electronic devices for the couriers, individualized routing for each package, based on destination, might be introduced into the system right at pickup.

Rather than simply addressing immediate problems, however, CEO Smith charged the developers of more advanced technology to aim for a system geared to the next century. Although collaboration with outside suppliers continued, more of the development shifted to within Federal Express itself. The result has been a sequence of increasingly sophisticated devices that gather added information useful for better serving customers, cutting down on potential errors, and increasing efficiency.

First, Supertrackers that could provide routing data right at pickup replaced the older Trackers. Once the bar-coded identification number pasted on a package is scanned, and the courier, prompted by queries in a tiny display window, enters destination, type of service, and other data, the Supertracker is plugged into the DADS terminal on the van. Software assigns routing data indicating where the parcel is to go for sorting—either the superhub in Memphis or to a regional hub—and this information is immediately transmitted to the local station's computer via microwave radio link. The local computer, in turn, enters the package data into the COSMOS IIB data base. Supertracker software also provides some checks on the accuracy of the data, such as making sure a proper zip code has been entered—perhaps preventing a mix-up between Richmond, Virginia, and Richmond, Indiana, for example.

Every package is scanned six times on its journey from pickup to delivery, each time updating COSMOS IIB so that tracking information is instantly available to customer service representatives. Supertrackers throughout the system feed more than 14 million transactions a day to an IBM 3090/600J computer in Memphis. The Supertrackers are plugged into racks when the couriers return from their day's rounds, and any changes in desired routing patterns can be fed into the devices for use with the next day's pickups.

In the fall of 1991, Federal Express began to put Astra printers into the vans, allowing couriers to print out bar-code Astralabels indicating routing, type of service, and so on. Federal Express even developed the glue that tacks these labels to packages because analysis of some missed deliveries pinpointed lost labels as one root cause. Use of clearly printed labels means that couriers no longer have to use magic markers to jot a universal routing and sorting aid (URSA) number on each package, eliminating another potential cause of errors.

Astralabels also go onto containers that consolidate many packages for the same destination, so that a single scan at transfer stations gives COSMOS updated information on all the nested parcels at once. As Astralabels are scanned at hubs during sorting, a "beep" sound tells the loader that all is well, but "beep . . . beep . . . beep" means a missort, which can be corrected before the parcel or container is sent on.

The additional data from what Federal Express calls the Astra Zodiac system require the expansion of transaction-handling capability at Memphis headquarters. Yet efforts at even further improvements continue. A more advanced portable scanner/computer for couriers, called the Megatracker, is under development.

Progress in tracking systems has gone steadily forward in spite of numerous hurdles that sometimes threatened to sidetrack or even abort major projects. Initially, Supertracker input was to be handled by a planned ZapMail system. To prepare for this possibility, transaction handling was completely reprogrammed from Digital Equipment to Tandem computers. Then the ZapMail network was scrapped. Plans to launch a dedicated Federal Express communications satellite were set back by the Challenger disaster. Again, the system had to be completely reprogrammed for the company's central IBM computer network. Without top management's vision for custodial tracking, it would have been difficult to get the support and budgets needed to keep moving onward in spite of such setbacks.

It was not just continuing technical development that required backing; there was also the problem of getting a fast-growing organization to adopt new ways of doing things. In 1987, for example, professional trainers fanned out across the nation to teach some 20,000 couriers how to use the new Supertrackers.

Federal Express continually looks to new technology to help it better serve shippers, but it also encourages employees to look for ways to increase service to their internal customers. In data centers, high priority is given to decision support and analysis aimed at streamlining operations, according to Ed White, vice president, computer operations. With more than 100,000 computers, including Supertrackers, DADS terminals, and desktop machines, the company has more than one computer per employee. This support has contributed strongly to continuous improvement efforts.

One analyst, for example, decided on his own to computer-simulate the operations of the Memphis superhub, where about a million packages are transferred among some 100 aircraft each night. Once the simulation model was running, he found that juggling the order of work at three loading docks could speed up departures by 22 minutes, a tremendous advance for the total process.

Since call centers must feed pickup requests to regional dispatch stations, it became apparent that data center computer power might help in figuring optimum routing. Thus, when pickup requirements are displayed for regional dispatchers, a recommended route is also shown. The dispatchers accept these recommendations more than 80% of the time and retransmit them to DADS terminals on the vans. But if they know of some special circumstance or problem, such as a traffic jam, they can manually override the system and switch the assignment to a different van's DADS computer.

The 16 domestic call centers receive about 250,000 calls a day, half of them for package pickups. If lines are busy, calls are automatically switched to another center to be answered in five rings or less. Training for agents includes computer-based self-learning programs. Employees must attain final scores of at least 85 at the end of 5 weeks of training in order to qualify to handle customer calls; otherwise they are assigned different jobs. For international calls, agents must be able to explain customs and any special licensing requirements. Again, technology provides assistance. Any changes in the rules are displayed for them, such as a refusal by the Philippines, for example, to accept any more packages with no dollar value assigned for contents.

Federal Express makes computers available to frequent customers under the Powership program, allowing such customers to tap right into COSMOS to check on their shipments and to print out invoices at the end of the day, simplifying billing.

Although individuals are encouraged to take the initiative and to contribute improvement ideas, changes in work processes are normally developed by quality action teams, which include employees who work in the particular area of concern. Customer surveys help guide improvement efforts. When it was discovered, for example, that many customers did not require morning delivery for some packages,

Federal Express instituted a separate, lower-cost class of service for delivery later in the day. In seeking opportunities for improved service to internal customers, the company suggests that good working relationships come from getting answers to three key questions:

1. What do you need from me?
2. What do you do with what I give you?
3. Are there any gaps between what I give you and what you need?

Every quarter, a dozen of the best teams from around the world come to Memphis to make presentations on their efforts to top management, according to C. Patrick Galvin, vice president, corporate systems. At one meeting, he reports, two teams came in from Italy and another team from Singapore, all supported by the Quality Department budget.

Unlike many companies, where a favorite expression is "that's the way we've always done things here," there is a sense at Federal Express that the way things are done will inevitably change, usually sooner rather than later. Driving this constant improvement is a multilevel planning cycle. Corporate management develops a 5–10 year vision, including such longer-range matters as new services, new airports, and new types of aircraft. These plans are worked out in more detail for a 2-5 year planning horizon. Yearly plans then evolve for such matters as schedules, crew requirements, and procurement of specific aircraft and ground vehicles. The quality push began with yearly goals, which soon became quarterly and then moved toward monthly targets.

With the SQI index and FXTV to communicate goals and results, everyone in the company is acutely aware of progress being made toward tough, well-defined targets. All bonuses in the company are dependent on reaching the goals that have been set.

Quality vice president Roberts likes to sum up the Federal Express approach with a quote from Will Rogers: "Even if you're on the right track, you'll get run over if you just sit there."

Chapter 9 ⤳

IBM Leads Computer Industry's Drive for Quality

For decades, U.S. companies reigned supreme in world computer markets. IBM gained a strong leadership position in every market it entered in Europe and Asia, let alone in the United States. It wasn't until the 1980s that Japan's vertically integrated electronics giants began to make inroads. In mainframes, investments in software and training effectively locked customers into a particular vendor's proprietary systems, ensuring a long-term user base for the American computer firms though a number of system generations.

It was easy to become complacent under such market conditions, particularly as annual sales and profits rose at healthy rates. But computer vendors now recognize that those good old days are gone forever. Along came the personal computer market, which was soon rife with low-priced clones of industry-standard computer architectures. More and more companies networked their desktop machines, and managers began to clamor not just for easier interactivity among the personal computers, but also for better access to central data bases. In mainframe and minicomputer segments pressure from users and user groups forced growing acceptance of international standards. Powerful users began to demand open systems so that an assortment of hardware platforms and peripherals could be interlinked, and data interchanged among software packages.

Even the most entrenched computer vendors have been forced to react to this shift in the markets. Not only must the companies be more attentive to traditional competitors, they must also face off against powerful new overseas competition, particularly such Japanese giants as Fujitsu, Hitachi, Toshiba, and NEC. Another threat comes from a new generation of competitors offering alternative system solutions, such as client-server networks and parallel processing systems.

With such heated competition, it's not surprising that computer firms have eagerly joined the total quality brigade. Total quality management offers computer makers a foundation for the rapid, customer-oriented market response they urgently need to meet global competition. This chapter, along with Chapter 13, presents details of the individual quality approaches being developed by a range of computer vendors.

IBM not only is the industry leader, but in 1990 its Rochester Division in Minnesota, which builds midrange systems and disk drives, was also the first computer manufacturer to win a Baldrige Award. First the corporate drive for what IBM calls "market-driven quality" will be described, followed by a more detailed discussion of the Baldrige Award–winning efforts of IBM's Rochester Division.

～ IBM: The Drive for Market-Driven Quality

When IBM-Rochester won the Baldrige Award, it boosted spirits throughout the world's biggest computer company. IBM had run into troubles in recent times, and winning the coveted Baldrige trophy was a sign that the direction set by top management for a strong recovery was the correct one. The road they chose leads to total quality, called *market-driven quality* (MDQ) at IBM.

Since taking the helm in 1986, IBM Chairman John Akers has been restructuring the $62 billion company in an attempt to restore its image and regain its preeminence as a technology and product leader. So far, results have been mixed at best. Recently, however, Akers has redoubled his efforts with the focus on quality.

Most of Akers's earlier efforts were in the people department. He removed two layers of management, cut the U.S. work force by 37,000—including 7,000 managers—and increased IBM's field force of sales and service personnel by nearly 20%. In the fourth quarter of 1989, IBM wrote off $2.4 billion in job-reduction and plant-closing costs. Even further cutbacks followed.

More distressing to IBM executives was the decline in market share. Since 1986, the company has lost market share among major customers in both mainframes and personal computers and will probably keep losing market share in software and peripherals, according to the Gartner Group, a research firm in Stamford, Connecticut "IBM is trying to come back, but it's a tough fight," notes Randall Brophy, a Gartner senior research analyst.

Big Blue was plagued by manufacturing problems as well. In 1989, difficulties with the production of the high-end PS/2 personal computer line delayed sales, while semiconductor snags slowed mainframe production. Manufacturing glitches also held up production of the recently introduced 3390 disk drive, while the RS/6000 workstation was delayed because of reported problems with the operating system. "Our business as a whole did not live up to our expectations," Akers told stockholders at the company's annual meeting in 1990.

In response, IBM's chairman launched a full frontal assault on what he now believes is the root cause of the company's problems: poor quality. At an IBM senior management meeting, Akers rolled out his heavy artillery: the market-driven quality program, or MDQ. The objective is to make IBM more customer-responsive and more productive—two changes management hopes will strengthen sales and boost profits.

The MDQ program has three components: a set of quality initiatives, a system of process review, and a system of quality measurement. All are aimed at cutting defects to near zero in everything IBM does and at shortening product cycle time. Beginning in 1990, MDQ affects every customer, every vendor, and every IBM employee worldwide. "This really is a survival issue," Akers told his senior managers. Echoed Jack Kuehler, IBM's president, "The hard fact is that we're not doing well enough."

Five-year figures compiled by *Electronic Business* magazine show Big Blue's revenue growth at 6.4% and net income growth at a negative 10.6%. Because of this track record, senior executives are serious about improving both product quality and customer satisfaction. After all, it is high-profile, quality-conscious competitors, like Compaq Computer and Hewlett-Packard, that have been stealing market share from IBM in recent years. "Every year the customer sets higher and higher standards," says Stephen Schwartz, IBM's senior vice president for market-driven quality. "If they don't get the quality they need from us, they will get it from someone else."

Schwartz moved into the new position in April, 1990, from the top job at IBM's Application Business Systems (ABS) division in Rochester, Minnesota, the 1990 Baldrige Award winner.

Although IBM won't say how much it invests in quality, Schwartz admits the cost will be high. Just training IBM's managers for 2 days requires about 800,000 hours, or at least $20 million. "It's costing us some money, but we are already seeing a payback," says Schwartz.

To drive home the MDQ message, Akers and his management committee— Schwartz, Kuehler, and a few other senior vice presidents—toured all of IBM's major U.S. facilities and many overseas sites. Akers sat down with the rank and file to impress upon them that MDQ is *the* top priority at the highest levels of the company. "Market-driven quality is the first thing on our agenda at every site," says Akers.

∼ Satisfying Customers Must Become an Obsession

Employee empowerment is an important quality theme, and IBM's chairman emphasized that message during a recent speech. "Market-driven quality starts with making customer satisfaction an obsession and empowering our people to use their creative energy to satisfy and delight their customers," Akers said.

Employees are pleasantly surprised by Akers' recommendations. "He told us to take more risks and spend less time double-checking everything we do," says Jim McDonald, manager of systems assurance at IBM's Entry Systems Division in Boca Raton, Florida. "I liked that."

What drove IBM's chairman to take such a radical course? After all, MDQ involves a sweeping cultural change at Big Blue. For one thing, during the 1980s IBM gained a reputation as a company that often ignored customers' demands. Today, improved customer satisfaction is seen as absolutely vital to IBM's future.

The argument for implementing a customer-focused corporate quality pro-gram gained momentum as the result of successes at the ABS division in Rochester. The division, which was a Baldrige finalist in 1989 before it won the award in 1990, manufactures AS/400, S/36, and S/38 midrange computers and data-storage devices. Since the early 1980s, the division also has run a quality-improvement program that contributed strongly to the AS/400 midrange computer line and other new products.

The AS/400 was an important factor in the corporate adoption of MDQ. In developing the computer, management broke with IBM tradition by involving customers at the earliest stage of product definition. The reason was that the AS/400 was to replace the S/36 and S/38 models already owned by 300,000 customers. The AS/400 program also was a testing ground for shortening product cycle: 28 months for the AS/400 compared with 5 years for the S/38, says Larry Osterwise, ABS director and site general manager.

Akers's commitment to quality got another push when a group of 31 top IBM executives visited Motorola, Inc., the winner of the 1988 Baldrige Award, for a 2 1/2 day session on quality. The delegation was headed by Heinz Fridrich, vice president of manufacturing, Bob Talbot, formerly director of quality and now assistant general manager for special bids, and Bob Friesen, vice president of IBM U.S.'s development and manufacturing. Back at IBM, the group urged top management to adopt quality initiatives.

They succeeded. But Akers's chances of pulling off a complete corporate cultural change is tempered by the fact that he is due to retire at age 60 at the end of 1994. Whether or not such a mammoth turnaround is possible with such a short tenure is unclear. Still, Akers says he wants "MDQ installed throughout IBM to the greatest degree possible" by the time he retires. Schwartz concurs: "MDQ is part of the legacy John wants to leave."

∽ No Easy Targets for Improvement

Akers does not believe in setting easy targets. He expects IBM's entire worldwide organization to slash defects by a mind-boggling factor of 20,000 over the next 5 years and cut average product cycle time in half. One of the central initiatives IBM has adopted is Motorola's "six-sigma" approach to eliminating defects. Six-sigma is a statistical term denoting about 3.4 defects per million operations (see Chapter 4 on Motorola for more details).

Kuehler nicknamed IBM's defect reduction initiative "excellence in execution." Whatever name the program takes, IBM's quality executives clearly have their work cut out for them. The corporate goal is to reach six-sigma by 1994. In January 1990, the company ran at an average level of three-sigma (66,800 defects per million operations), according to Kuehler. Most experts rate U.S. manufacturing on the whole at less than four-sigma.

Achieving six-sigma within 5 years will involve a sea change throughout IBM. Incremental goals were set at a number of divisions. One division, for example, plans a 10-fold reduction in defects each year through 1993 and a 20-fold reduction in 1994.

While a number of senior manufacturing and marketing executives are optimistic about their goals, many doubt that six-sigma is achievable by 1994. Even Schwartz is hesitant when asked if IBM will make its six-sigma deadline. "Some say yes, but time will tell," he says. He prefers to stress the positive: "If you don't set ambitious goals, then you don't change the thinking of the people."

Cycle time reduction is IBM's other primary concern. IBM wants to compress the process that starts when a customer expresses a need and ends when a customer pays the bill. That includes everything from revamping intelligence-gathering—"market information capture," as IBM calls it—to speeding up design, development, manufacture, ordering, shipping, and billing.

IBM places much emphasis on intelligence gathering. Without good market information, designs are constantly in flux, and product development is often delayed. "There are a lot of tools for shortening the cycle, but they work only if the requirements don't change," remarks Schwartz. The AS/400 is a prime example of using customer input to shorten the product cycle, he says.

To bolster information gathering, IBM is combining its disparate market research data bases into one, so that consistent and complete market information can be disseminated to IBM employees around the world. There is no schedule for merging the data bases, but Schwartz estimates that it will take several years.

Meanwhile, managers are receiving the tools to make six-sigma happen. By mid-1991, all 48,000 managers worldwide had attended an MDQ training session to equip them to adapt the material to their specific needs and teach their subordinates. Akers and his inner circle of senior executives had already received training.

Management training involves a 2-day MDQ session and is conducted by IBM staff, including top executives such as Schwartz. Both Akers and Terry Lautenbach, senior vice president and IBM U.S. general manager, make video presentations. There also are contributions from outsiders, such as Paul Noakes, vice president and director of external quality programs at Motorola, who discuss their companies' quality.

The 2-day course includes most of the quality basics—defining initiatives, deploying a quality-based process-management system, applying the Baldrige criteria, implementing six-sigma, benchmarking, and measuring quality. These managers then take training material back to their employees.

Market-driven quality also applies to nonmanufacturing sectors of the company. Ken Thornton, general manager of marketing in the mid-Atlantic region, attended a managers' MDQ session and quickly set up a quality program. All 7,400 of his employees have now received MDQ training. Sweeping changes already have occurred, notes Thornton. Each branch office is required to set five measurable quality parameters for improving customer service. The parameters can be as simple as the time it takes to return a phone call or as complex as improving employee skills. To stress his commitment, Thornton appointed a regional MDQ manager responsible for implementing MDQ, who reports directly to him.

In all branch offices in the mid-Atlantic region, Thornton installed a new suggestion system. Within the first 3 weeks, the Hagerstown, Maryland, branch was deluged with 128 suggestions from its 54 employees. The spectrum of ideas was broad. One employee recommended the accounts-payable staff visit customer sites to solve certain problems, while another looked to improve the efficiency of the branch library, says Howard Rockwell, branch manager and 37-year IBM veteran. Many suggestions have already been implemented, he adds.

At IBM's Santa Teresa Laboratory for the development of software, in San Jose, California, management uses Baldrige criteria to measure each function within the division, says Tom Furey, assistant general manager of programming systems and site general manager. Measurements include information gathering, customer satisfaction measurements, and work force use for all 2,000 employees. In one product development group, a quarterly score was a disappointing 390 out of a possible 1,000. After some process changes, says Furey, the score improved to 570 by the following quarter.

Making MDQ work—and in particular, making six-sigma a reality—must

involve suppliers. But Schwartz is still unsure about how much pressure will be necessary to convince suppliers to adopt their own MDQ principles. "We will ensure that
quality is up to scratch, but at this time there is no mandatory MDQ supplier program," he says. For now, the pressure is relatively subtle. Over the past few years, IBM
has been paring its supplier base and establishing closer relationships with vendors, or
"business partners" as the company calls them. The number of IBM's significant suppliers dropped from 4,000 in 1988 to 3,000 in 1990, due in part to higher quality standards. Given the threat of further cuts, the pressure on suppliers to perform is sure to
drive up vendor quality.

One major IBM supplier has already responded. National Semiconductor
Corporation in Santa Clara, California, recently unveiled its own quality initiative,
which is compatible with most of IBM's goals, claims Tim Thorsteinson, National
Semiconductor's director of quality performance. Thorsteinson is responding to the
needs of all its customers, not just IBM. "I think all computer vendors are redoubling
their efforts, including IBM," he says.

The task could demand more than subtle pressure. At IBM's Rochester facility,
general manager Larry Osterwise says he *demands* supplier compliance in terms of
focusing on quality, understanding and implementing the Baldrige application criteria, and sharing suppliers' quality approaches with IBM. The alternative, he says, is to
lose IBM as a customer. One West Coast supplier, who asked not to be identified, confirmed IBM's demands, saying that Big Blue requires adoption of both the Baldrige
application criteria *and* six-sigma. "IBM may have the most extensive quality requirements in the industry," he says.

While MDQ is still in the early stages, there is little doubt that as it progresses
in other IBM divisions around the world, they too will be forced to place more and
more stringent demands on suppliers. "We expect them to strive for constant improvement," says Schwartz.

ᓇ *The Three Legs of Market-Driven Quality*

IBM's quality program is an amalgam of measures borrowed or adapted from
other companies. "Whatever works, we use," says Stephen Schwartz, senior vice president for MDQ. Still, the framework has been honed to fit IBM's needs. The intent is
to take the message to all 380,000 IBM employees around the world, starting with
managers. Each manager—defined as anyone with responsibility for people—will be
charged with learning the basics of MDQ and adapting the program to fit departmental needs.

In essence, MDQ rests on three legs:

1. *Initiatives:* Initiatives are five points that encompass the market-driven
   principles (define market needs, eliminate defects, reduce product cycle time,
   measure progress and commitment, and empower employees;
2. *Process Review:* Process Review is a system for analyzing the flow of business
   and making changes where necessary;
3. *Self-examination:* Self-examination includes methods for measuring and
   comparing IBM's quality over time against internal and external standards.

In all respects, IBM is aiming for a new way to do things. A defect-reduction goal, derived from Motorola's six-sigma program, is to reach 3.4 defects per million operations by 1994. In January 1990, IBM had an average defect rate of about 66,800 defects per million, or three-sigma. The goal is quite ambitious. Six-sigma demands a whole new way of looking at all aspects of the business, from designing mainframes to answering telephones.

As for review, IBM is intent on doing things the *right* way, not just the IBM way. Formal processes are being instituted that will allow continuous quality improvement and cycle time reduction.

In the area of self-examination, IBM has instituted its own quality award, which is based on the Baldrige Award criteria. The goal is to create internal competition among IBM entities and bring all divisions up to the same standard.

In 1990, two IBM divisions entered the Baldrige competition. One was second-time entrant Application Business Systems division in Rochester, Minnesota, which won an award, and IBM Credit Corporation, which was unsuccessful. Says John Akers, IBM's chairman: "Winning is nice but competing for the Baldrige Award is more important. The value is in the process."

⌒ *IBM-Rochester: Striving for a Perfect "10"*

When Olympic gymnast Peter Vidmar visited IBM's Rochester, Minnesota, plant, he told the 8,000 employees that performing well in a gymnastic routine rates about 9.2 points out of a possible 10.0. Technical excellence adds 0.2 to the score, risk taking another 0.2, and innovation 0.2 more. To rate a perfect 10, Vidmar added, the gymnast must earn a final 0.2 points for "virtuosity."

His audience got the point. Ever since IBM chairman John Akers kicked off the company's market-driven quality (MDQ) program, IBM employees have known that their own virtuosity is rated every day. Although the Rochester unit—a producer of midrange computer systems and data-storage products—hasn't yet scored a 10, improvement efforts at the site serve as a model for IBM's sweeping new drive for total corporate quality.

After an unsuccessful bid in 1989 for the Malcolm Baldrige National Quality Award, IBM-Rochester brought home the bacon in the 1990 competition. The plant is upgrading nearly every one of its operations, with the emphasis on customer satisfaction. Product design procedures are being streamlined, product cycles shortened, inventories trimmed, and waste and rework reduced.

The Rochester plant's philosophy can be summed up in seven words, says Larry Osterwise, site general manager and director of IBM's Application Business Systems group: "If it's not perfect, make it better." Osterwise adds that this is closer to the Eastern philosophical view now widespread in Asian industry than it is to the usual attitude of American industry, usually expressed as: "If it ain't broke, don't fix it."

IBM-Rochester, like so many other Baldrige contenders, learned that applying for the award is itself an exhausting exercise. In this case, the process required some 20 person-years of effort, according to Roy Bauer, manager of engineering planning and operations at Rochester. Bauer, who coordinated the work on the winning entry, and his team uncovered quality problems at practically every turn. The result was 71 "wart reports" that were divided into 34 top-priority tasks and 37 less urgent problems.

"We're working on problems other companies don't even know they have!" says Bauer. People were asked to develop strategies and action plans to work on each problem. Rework levels were too high in some production operations, for example, so the division made efforts to gain better control.

Rather than going for exotic, complex solutions, the quality team concentrated on "basic blocking and tackling," says Osterwise. That requires coaching, and Rochester managers often exchange visits with other U.S. companies known for their strong quality efforts. David Kearns, who was CEO of Xerox when it won an early Baldrige Award, spent 6 hours touring the Rochester facility. IBM employees themselves have visited such innovative companies as 3M, American Express, and Disney.

"Although Disney is in the entertainment business, their quality procedures were not that different from ours," comments Richard Lueck, Rochester's director of site services. Lueck stresses the importance of a wide range of benchmarking visits because even the best organizations may excel in only some areas. American Express appears to be the best in the business in some billing and accounting functions, for example, Lueck says, but IBM has some 14 different financial groups, and it's hard to find good models for each of these specialties. Since CEO Akers stresses customer satisfaction, many organizations in a variety of businesses were benchmarked, according to Lueck, generating many good ideas even though no single organization matched the scope of IBM's own service operations.

With the help of such insights, plus a wealth of suggestions from IBM Rochester's own people, the customer satisfaction campaign is reportedly paying off. The effort has been particularly strong in the software sector. Even the name of Rochester's Software Development Support Center was changed to Software Partners Lab in order to stress the importance of customer participation. Customer teams come to the lab from a wide assortment of businesses—financial services, oil exploration, workstations, health maintenance, even bowling-alley construction—and work with IBM employees in well-equipped offices to perfect new applications.

Largely because of the Baldrige competition's heavy emphasis on customer satisfaction, Rochester constantly pushes its involvement with buyers upstream. Before the AS/400 midrange computer was introduced in mid-1988, customers were asked to join round table talks in which IBM discussed its plans, laid out various options, and encouraged customers to help set priorities. They discussed more than 155 key issues related to attaching PCs to the new IBM systems, and, of these, identified 14 as high priority. Acting on the customers' recommendations, IBM agreed to share software codes with other manufacturers of equipment that was used by the customers, according to Ray Harney, manager of Software Partners Lab.

The Rochester plant also worked to speed products to market. While the development cycle for the S/38 took a full 5 years to complete, the cycle for the AS/400 was cut to just 28 months by using cross-functional teams that merged software and hardware design with manufacturing. IBM has called the introduction of the AS/400 the most successful product launch in its history. Within 6 months of introduction, more than 25,000 AS/400s were installed worldwide, with two-thirds of those sales overseas, according to Osterwise. Exports from the Rochester facility were estimated to total $575 million in 1989 alone.

Customer input shaped the fundamental architecture of the AS/400 family. Because of the popularity of object-oriented programming, operating system software was geared to running this type of code. To get still more information, IBM shipped thousands of machines early for evaluation by customers.

The customer link does not stop once a system is shipped. After 90 days, a "customer partner" telephone call is made to check on any problems with the new installation. IBM retirees, working part time, make many of the calls because of their familiarity with the company and its product line. "These are not survey calls," says Robert Tremain, a market-support representative. "We train our people to listen." Research and development staffers also get a chance to participate in these calls, reports Osterwise, in order to receive direct feedback from users.

Customer problems are referred to the local servicing group for action, and then a 30-day follow-up ensures that the problem has been solved. If customer complaints happen to come through chairman Akers's office, company policy requires a response within 10 days, according to Margie Spohn, manager of customer partnerships and telemarketing. What if the field-service representative can't handle the problem or doesn't have the right parts? "Someone gets on an airplane," Spohn says. "Surveys show that we have less than 5% dissatisfied customers, and that is going down."

IBM-Rochester favors continuous-flow manufacturing and just-in-time methods to shorten production cycles and reduce inventories. IBM's management studied the techniques of Shigeo Shingo, the Japanese consultant who helped set up Toyota Motor Corporation's system, but they decided to take a more gradual approach toward just-in-time deliveries, says Osterwise. This is important because a glitch in the production line can mean missed deliveries to customers. He likens the approach to guiding a large ship through a narrow channel with many submerged rocks while slowly lowering the water level. This will reveal a few rocks, and then a few more, with the course being altered each time to avoid any new hazards.

Indeed, one thin-film disk drive line in the plant was shut down during a tour of the facilities because of a shortage that had resulted from operating with steadily lower inventories from a Guadalajara, Mexico, plant—the source for actuators to go into the drives. Later the reason for the shortage was located and eliminated, enabling inventories to be lowered even further.

To provide flexibility on the disk-drive production floor, strips of colored tape designate locations for the rolling five-layer carts used to deliver parts and subassemblies to workstations. This is known as a *kanban*—or pull-when-needed—system. Blue-taped areas indicate items in preparation for production; green tape identifies disk drives that have gone through testing; and black-taped areas are rework stations.

Efforts to empower workers are evident throughout the disk-drive production area. Lists of worker committees, dealing with issues that range from safety to six-sigma quality, are posted in the office of George Thompson, manager of Rochester storage products operations. The worker committees do produce results. Thompson explains that incremental goals for a new disk drive had been exceeded well before deadlines. The initial move to 4.0-sigma—the equivalent of 6,210 defects per million—from 3.85-sigma happened so quickly that a new goal of 4.5-sigma had to be substituted, and that, in turn, was within reach nearly 2 months early.

Cutting down on even rare slip-ups requires every worker to thoroughly understand each work process. To help prevent mistakes, posters labeled Hot Buttons, with flames drawn in to call attention to the instructions printed on them, are hung above some work stations in the disk drive assembly area, one poster reads:

> PROBLEM: Oil in the bearings.
> SOLUTION: 1. Don't put oil in the bearings.
>   2. Put oil in the sleeve.

Before beginning to work on the disk drive line, a new worker is given a guided tour of each station by an experienced trainer. The following week the same procedure is repeated, but this time with a new trainer, just in case some vital point was missed the first time through.

Osterwise stresses the need for creating an environment of high morale so employees want to do close to perfect work. Improvements must come both incrementally and in quantum leaps, he says; only then will virtuosity be achievable and the perfect 10 be possible.

~ Baldrige Award to the Winner of IBM's Own Competition

Self-assessment is a critical part of IBM's quality journey. To drive the message home, IBM initiated its own quality award in 1989, using about 45% of the Malcolm Baldrige National Quality Award criteria in its evaluations. That year 62 of IBM's 102 divisions entered the in-house competition. Four awards were presented, one of which went to IBM-Rochester, a finalist in 1989's Baldrige competition. In 1990 the Rochester division was a winner of the Baldrige Award.

Housed in 35 buildings on nearly 600 acres, the Rochester facility accounts for only 2% of IBM's work force but 8.5% of its total revenue. Aside from manufacturing midrange systems and disk storage devices, the facility also provides support services ranging from research and development to customer follow-up.

Chapter 10 〜

How Baldrige Winners and Contenders Boost Factory Quality

Unlike most races, in the Baldrige competition losing does not mean failure. Baldrige Award candidates stoutly maintain that it isn't the trophy that counts but the competition itself: the rigorous, sometimes painful, always revealing process of learning what it takes to become a world-class competitor. That's why the experiences of four companies that have made dramatic moves to instill total quality concepts into their factories should provide ideas useful to any manufacturer. All four of the companies profiled here failed to win the Baldrige competition, some more than once. Following unsuccessful attempts, however, two of them responded so well to the feedback provided by Baldrige examiners that they subsequently became winners. The other two are strong contenders with hopes of winning the award in the future.

The organizers of the Baldrige program hoped that striving for a national quality award would help keep American companies competitive in global markets. In manufacturing particularly, U.S. firms have been attracted offshore to take advantage of lower wage rates and, in some cases, less stringent regulations. Many observers have expressed fears that, especially in high-growth advanced technology sectors, this could lead to a hollowing out of American industry. U.S. companies could become primarily marketing outlets for products made with foreign components and assembled abroad. The long-term effects could be devastating to the U.S. economy. Aside from millions of jobs, critical skills, not just in production but also in research and development, could eventually be lost. This has already happened in the consumer electronics sector. First U.S.-owned domestic manufacturing and, then, research and specialized technology capabilities in a once-thriving but highly competitive U.S. industry almost disappeared.

It already appears that the quality push by American manufacturers has had an effect, and that more production may move back onshore. The first company described here illustrates the beginnings of such a reverse trend. Solectron, a 1991 Baldrige Award winner, is a rapidly growing assembler of computers and subsystems for such major U.S. vendors as IBM, Apple, and Hewlett-Packard. Excellence in manufacturing is indeed encouraging large assembly firms either to return or to keep production in the United States. By serving several vendors, contract manufacturers can afford to use the latest production technology. Communications and coordination are simplified compared with offshore factories, and contract manufacturers can collaborate with the assembly firms' designers to provide rapid ramp-up to production. This is an important advantage, as tougher global competition demands shorter time-to-market cycles. By using American workers, the assembly firm does not transfer its manufacturing know-how abroad.

Control Data's case shows the value of the focused factory concept. Cost savings from moving some manufacturing to lower wage rate areas can be eaten up not just by difficulties in coordinating production but also by duplication of both physical space and administrative support. Speedier and better coordinated assembly in a focused factory can allow key assembly decisions to be left until much later in the production cycle, Control Data found. This allows much quicker response to customers' needs, a vital part of total quality management, and can bring in sales that would have been lost due to the delays and snafus inherent in long supply and communications lines. How many calculations of cost savings from offshore manufacturing also take into account potential quality problems and the opportunity costs of lost sales?

Perkin-Elmer demonstrates how an integrated approach to total quality management, starting with the design process and extending not just into manufacturing but also into cost accounting methods, extends the range of opportunities for continuous improvement within the context of overall corporate strategies.

Zytec's strenuous efforts to follow the precepts of Deming and others show that even small firms in tough markets can build a strong competitive niche by concentrating on total quality methods in the manufacturing process. In Zytec's case, the process led to a 1991 Baldrige Award as well.

∿ Solectron: CEO Chen Crusades for U.S. Manufacturing Excellence

At a meeting of some 900 suppliers with one of the largest U.S. electronics corporations, the president and chief executive had just finished an address on quality and asked if there were any questions. "Yes, I have a question," called out Winston Chen, president of Solectron Corporation, a San Jose contract manufacturing company. "If your company is so dedicated to quality, then why do you pay your marketing executives 25% more than your manufacturing executives?"

The same question could have been asked of just about any one of America's corporate giants, but Chen was challenging Motorola, a 1988 winner of the Malcolm Baldrige National Quality Award. Such irreverence is a way of life with the former IBM researcher who now runs the third largest company in the fast-paced electronics contract-manufacturing industry.

After failing to win the Baldrige Award in earlier attempts, Solectron itself became a winner in 1991. With more than 2,000 employees, Solectron had to contend with the bulwarks of American industry under the Baldrige rules, since it easily surpasses the 500-employee limit set for the award's small-company category. Still, with a growth rate well over 25% and sales that passed $265 million in 1991, Solectron is typical of the successful small, high-growth companies so common to the electronics and computer business.

One reason Solectron decided to compete was that it launched its quality program in 1982, well before the Baldrige Award came along, mainly because of Chen's views on how American industry needed to mend its ways to match powerful global competition. The stringent Baldrige criteria now provide standards against which smaller companies such as Solectron can test their progress toward world-class manufacturing right along with such U.S. giants as IBM, Motorola, and Xerox, which have far greater resources and large corporate quality staffs.

Early failures to win a Baldrige Award in 1989 and 1990 did not dissuade Chen in his crusade for total quality. Indeed, he remains a proselytizer of the new faith sweeping U.S. corporate culture. "Only by reestablishing leadership in manufacturing excellence can the United States hope to maintain its standard of living and regain its place in the world economy," says Chen. He sees the national Baldrige Award competition as an excellent focal point for stepping up the global competitiveness of U.S. manufacturing, and Solectron openly discusses its own efforts. But he believes that nothing short of a major crusade to reestablish the once-proud preeminence of U.S. factories can reverse this nation's decline. Doing this, in Chen's view, will take a major shift in the thinking of corporate management in the United States.

"In Japan the differential between the factory worker's salary and top management's is about 6 to 1; in the United States, it is much higher, and (with bonuses and stock options) can go up to 600 to 1!" Chen points out.

How can workers accept management's preaching about how vital their contribution is on the factory floor when their paychecks send so different a message, he complains. Chen urges American industry to greatly reduce the currently unbalanced salary structure, which he feels overcompensates upper management. Justifying such large differentials also encourages organization charts with many layers of middle management, which tends to separate decision making from the manufacturing plant.

∼ Why Favor Marketing over Manufacturing?

The values of U.S. industry are also reflected in the disproportionate number of marketing and finance people versus manufacturing specialists who run large U.S. companies, Chen charges. In his view, overemphasis on financial manipulation, such as conglomeration and leveraged buyouts, along with the neglect of America's factories, has been the primary cause of our huge trade deficits and the decline of the U.S. manufacturing base.

Chen attributes recent improvements in the trade figures more to devaluation of the dollar, which also allows foreign buyers to grab up American assets at bargain-basement prices, than to a resurgence of U.S. factory productivity. To compete successfully in manufacturing industries, says Chen, U.S. management needs to blend the best methods developed by the Japanese with American innovation and advanced technology. That is what he is attempting to do within Solectron, a fast-growing contract manufacturing company.

Japanese methods adopted by Solectron include kaizen (continuous improvement by doing little things better and by setting and achieving ever-higher goals), total quality control, statistical process control, quality circles, just-in-time (JIT) manufacturing, total productive maintenance, *kanban* (the delivery to the next workstation or machine by request rather than pushing work along and maintaining buffer stocks), continuous-flow manufacturing, and the use of fishbone charts that display potential upstream causes of process variations that can affect quality. The company has been working on integrating these and similar techniques into its operations.

Chen believes American management should study the ideas of Japanese manufacturing gurus such as Taiichi Ohno, a Toyota vice president who helped the company achieve annual inventory turnover of some 90 times, versus the 3.5 times common in U.S. plants. Toyota accomplished this mainly through dramatic reductions in cycle times (particularly machine set-up times), by drastic cuts in inventories, and by greatly reducing lot sizes.

Cycle-time reduction and just-in-time manufacturing force better quality, Chen points out; otherwise, the production line would have to shut down too often. Shigeo Shingo, a consultant to many of Japan's leading manufacturers, writes about Ohno's techniques and also is noted for helping Toyota devise the "single-minute exchange of die" method that helped to vastly cut set-up times in automobile production. In addition, Shingo is known for the foolproof approach to design known as *poka-yoke*, which is not allowing the wrong button to be pushed, for example.

In one case at Solectron—the assembly of a complex mainframe disk drive with more than 2,000 components—the inventory needed was cut from $18 million to $3.5 million, cycle time was reduced from 10 days to 2 days, and the defect rate dropped from 100 to 2 parts per million. This was accomplished through a multi-pronged approach, says Chen. The techniques included

- design for manufacturability
- achieving zero defects through total quality control and statistical process control, and by using source inspection and Shingo's *poka-yoke* approach
- cutting cycle time with continuous-flow and just-in-time manufacturing methods
- using automation as appropriate
- setting up computer-integrated manufacturing

Chen is also an advocate of Taguchi methods. The Japanese quality specialist Genichi Taguchi devised methods of designing simplified manufacturing experiments to quickly determine near-optimal operating levels of complex systems. This is essential, Chen says, because probably 80% of variability in manufacturing is beyond the control of workers. At Nippon Denso, a Japanese auto parts manufacturer, for example, workers have designed more than 2,000 such experiments. Chen estimates that this company alone may have conducted as many Taguchi-type experiments as have been carried out in all of American industry combined.

"It is not only Japanese competition facing a domestic manufacturer," Chen points out. In Singapore, for example, an Economic Development Board ensures a favorable climate for foreign manufacturers. International telephone and telegraph rates are kept low, and wages for laborers averaged about $2 per hour in the late 1980s. In addition, the government offered a standard 30% manufacturing investment tax credit, whose counterpart was eliminated in the United States. For high-tech industries, the credit ranged up to 50%.

One result is that this island nation, with a population of 2.6 million, was able to create 50,000 jobs in the disk drive industry alone. Wages elsewhere in the Pacific Rim, such as in the Philippines or Thailand, range down to 40 cents per hour or less, although there may be a shortage of engineering skills in some of these nations. Because of such low wages combined with government nurturing of foreign manufacturing, the United States has been running a negative trade balance in high-tech electronics manufacturing since 1984.

Still, Chen believes U.S. onshore manufacturers can be competitive in many product areas by adopting Japanese techniques and blending them with the best U.S. practices. Supporting his contention is the steady growth of electronics contract manufacturing-assembly in the United States. Some $9 billion of production was outsourced to some 2,000 domestic contract manufacturing sites by U.S. original equipment manufacturers in 1989, out of a total U.S. printed-circuit-board assembly market estimated to be about $35 billion a year. The contract manufacturing industry has been growing 15% annually in the United States, with Solectron growing at an even faster 25% per year recently.

~ Up-to-Date Equipment Adds Speed, Agility

One reason that electronics systems firms increasingly turn to contract manufacturers is the availability of advanced production equipment. Solectron, for example, spent more than $20 million from 1985 to 1990 to set up 14 surface-mount technology lines, according to Chen, and has plans to add six more high-speed lines. As a result, it did some $90–$100 million in surface-mount technology business in 1990 for such clients as IBM, Apple Computer, Sun Microsystems, Conner Peripherals, and Hewlett-Packard.

Another major reason for systems firms to turn to contract manufacturing is to cut down the time needed to get a new design into production. Solectron sets up multidiscipline project management teams to work with customers on start-up projects. To speed the process, with help from all the project teams, new standards were generated for design review, documentation, and building prototypes. There is a clear focus on understanding the details and scope of the project and working in partnership with the customer to define requirements. Solectron carefully reviews the design with the customer with regard to manufacturability and reliability and passes on any savings achieved through such a review to the customer.

The company has been able to cut the cycle time for surface-mount technology boards from 15 to 5 days by using what it calls in-line manufacturing. The production equipment for each product, including test systems at the end of the line, are arranged in a straight line through the plant to speed product flow. This contrasts with many plants where employees perform each operation in a separate work area, and batches of boards move from one station to the next until they are completed.

While most statistical process control is still done by hand in Japanese factories, U.S. manufacturers have been much more adept at using computers for such tasks, Chen points out. In Solectron's plants, for example, laser scanners at each workstation read bar codes to indicate the circuit boards that are being stuffed or tested, as well as to identify the workers doing the job. When workers test soldered surface-mount-technology boards, they feed the results into a real-time computer system and analyze them using statistical process control. If they detect any out-of-control conditions, they flag the operator to stop the line. A computer simultaneously generates phone calls to the process and quality engineers and to the production supervisor, identifying the defect and the line on which it occurred. Since this system was implemented on the surface-mount-technology lines, under the direction of Peiman Amoukhteh, a production manager, defects dropped from 1,200 to 160 parts per million and continue to decline.

Even the use of flexible manufacturing for continuous production in Japan commonly allows only a choice among a number of options rather than complete flexibility, says Chen.

Training of workers is another important area for improvement in U.S. factories. At Solectron, new employees receive at least 40 hours of classroom instruction before going to work on the line. As a result of its training skills, Solectron received a $1.4 million grant from the state of California to teach about 600 factory workers and managers the techniques of statistical process control and materials requirements planning over an 18 month period.

～ U.S. Manufacturers Need to Stress Service

Focusing on customer service is another important approach to building an advantage for onshore manufacturing, Chen says. One goal at Solectron, for example, is to always answer a customer's phone call within four rings. In addition, the company checks the quality of the responses that the customers receive and feeds the results back to the telephone operators.

Solectron claims that its open communication policy is unique in the industry. It shares with customers the results of its quality audits, first-run yields, and Pareto charts (a sequence of bars showing the numbers of each type of defect in descending order). The company openly discusses its weaknesses along with plans for overcoming them.

With the help of such actions, the percentage of material returned fell from 2.2% to 0.4% in a little over a year, the company reports. Any returned boards are repaired and returned within 3 to 5 days, compared with 10 days before the effort to cut turnaround time was implemented in January 1988.

Anyone in the company may receive a customer complaint, but it must then be reported to the customer service representative and project manager. All such problems are discussed at a weekly meeting with the customer and then reviewed at a meeting with upper management. Customer satisfaction indices are reviewed along with any outstanding problems to be addressed. The grading system has been designed to emphasize any customer dissatisfaction.

"And one more thing," says Winston Chen. "We pay our manufacturing executives *more* than we pay our marketing executives."

∼ Control Data: Quick Response in a Product-Focused Factory

In the early 1980s, Control Data Corporation in Saint Paul, Minnesota, like many U.S. companies, had plants scattered around the country. Each factory made a portion of the company's computers, disk drives, lottery machines, and other products. But a push toward total quality management led to radical changes in manufacturing at Control Data. One result was the consolidation of eight scattered manufacturing facilities into a *product-focused* factory run by multidisciplinary teams in one large plant in Saint Paul. The move dramatically improved space requirements, product-cycle times, defect rates, and inventory turnover, according to George Ballata, general manager of computer systems manufacturing. Similar principles are being applied to the production of low-volume, highly complex computers in the Cyber 960 and 2000 lines, and higher-volume, simpler equipment (including intelligent wagering terminals, disk arrays, and networks) in another section of the same plant.

The product-focused factory, developed by Control Data in 1987, was a key element of its Computer System Manufacturing group's entry in the 1990 Baldrige competition. Although Control Data did not make the final cut, according to Ballata, the experience measured the operation's strengths and weaknesses against a recognized national standard and helped communicate the group's goals to the entire corporation.

Each production line at Control Data is run by a zero-defects team consisting of a quality coordinator—a permanent member of all teams—a process designer, a test technician, an operation manager, and several production operators. The teams are charged with satisfying customer requirements for quality products, meeting delivery dates, controlling costs, and responding quickly to market demand. To meet these needs, according to Ballata, the teams strive for steady improvement in what Control Data calls the five key measurements:

- internal and external yields
- cycle time
- total cost
- compliance to monthly targets
- inventory turnover

The system tracks both external yield, or the percentage of shipped products that are acceptable to customers, and internal yield, which is an accumulation of individual process yields. Since 1988, there has been a tenfold improvement in external yield, says Ballata, leading to major savings in the costs of testing and servicing systems. More big savings have come from manufacturing in one facility, which cut down space requirements to about a quarter of that once used throughout the scattered plants. Furthermore, computer orders that once took 12 weeks to fill are now turned out in 3 weeks. Final configuration, which is based on a customer's requirements for memory, input/output, cabinets, and so on, can now be held off until only 5 to 10 days before ship date, according to Ballata.

Training was a vital factor in achieving many of these gains. Ballata recalls that the performance of work teams did not pick up dramatically until after team members

went through courses in problem-solving dynamics and the whole production operation became self-directed. After the training, the teams devised new sets of rules. The
company instituted certification system for each task, and certificates now must be
posted at stations where workers are assigned. Supervisors had to abandon their traditional performance appraisals because, with the increased training and the new rules,
problems were quickly solved without the managers' assistance.

The new approach was evident at one workstation where an operator checked
on an apparent error detected by the robot used to check chips placed on a surface-
mount printed circuit board. In this case, the fault signal was cleared simply by adjust-
ing the optical system that laser scans each connection for defects. If that hadn't
resolved the problem, the operator immediately would have called the team's process
engineer. If the problem had threatened to disrupt production for 30 minutes or more,
the operator could have called a team meeting.

This procedure is covered by the "quickly" rule, one of a set of rules developed
by the zero-defects teams after going through problem-solving dynamics training.
Some of the other rules include

- the "actually" rule, which calls for the person who caused the defect to
  exactly replay the faulty operation in order to determine causes and to help
  others learn from the experience
- the "himself" rule, in which the person who caused or found the error fully
  explains what happened to the rest of the team
- the "don't speak" rule, which prohibits the manager from speaking first. Not
  only does this help team members to be less inhibited, but it also spotlights
  the workers' own responsibility to prevent future errors
- the "support" rule, which restrains managers from entering the discussion
  until the final stage of a meeting, and then calls for them to support team
  efforts to track the problem back to root causes, so future problems can be
  averted, rather than assessing blame or lecturing

As a further incentive to progress through cooperative efforts, Control Data
offers bonuses to work groups based on quarterly improvements in any of the five key
measurements in a program called success share. If a factory beats its annual budget
targets, 10% of the savings goes to the teams involved.

The product cycle for new computers in the Cyber 2000 line was greatly shortened by another team effort that brought manufacturing into the early design process.
Just one element of the resulting design greatly reduced assembly time, and reduced
costs, for the complex Cyber systems. The circuit boards in the new computers use
zero-insertion force connectors that link to two side panels rather than to the normal
back panel. These side panels are 18-layer circuit boards each with a buried resistive
layer for termination of emitter-coupled logic (ECL) lines (the computer uses about
6000 ECL gate array chips). This design eliminates the wirewrap back panel and external coaxial cables (since the signal terminations are now built into the circuit boards)
and thus greatly speeds assembly. Originally CDC made the complex circuit boards in
house, but this operation was sold off, thus further lowering costs because now, with
additional customers for the spun-off operation, there is greater utilization of the
advanced equipment required.

To keep in-process inventory down, a WIN (When It's Needed) wanding sta-tion is located in each work area. A laser wand connected to a personal computer reads bar codes on parts and subassemblies, ensuring resupply when needed based on mini-mum stocking levels worked out by material planners and manufacturing engineers. CDC reduced its supplier base from 850 in 1985 to 200, and it is electronically linking a growing number of vendors to the plant via electronic data interchange (EDI) so that immediate shipments can be made just when parts or materials are needed.

~ Perkin-Elmer: Managers Look Beyond the Bottom Line

Joseph E. Malandrakis has some pretty strong thoughts about what ails many American companies. "Management sets up systems that don't work, using the wrong designs and the wrong manufacturing methods, then measures results with the wrong yardsticks," he says bluntly. He backs up this candid view of what's wrong with much of American industry by pointing out that executives commonly demand simplified reports that leave out critical factors. Then they issue dictates to subordinates such as: "Factory utilization is poor. Fix it."

One reason for this state of affairs, suggests Malandrakis, a division vice presi-dent for Perkin-Elmer Corporation, based in Norwalk, Connecticut, is that managers' attention is too often glued to the bottom line. Traditional cost-analysis alone fails, he says, because too many critical portions of the operation are lumped into overhead, and vital contributing factors simply are not statistically measurable. Still, he agrees that progress toward goals must be tracked. Even with inadequate measurements, the company may achieve pockets of progress. Such efforts, however, are likely to have little success in reaching optimal solutions that extend globally over all operations.

Perkin-Elmer's quest for better management helped the producer of labora-tory instruments win Connecticut's own quality competition, which is based on Baldrige criteria. The company also was a strong contender in the 1990 Baldrige com-petition.

To better tie total-quality efforts to corporate goals, Perkin-Elmer begins with a vision that encapsulates management's worldwide strategic objectives, including asset management, product development, and quality performance. The company then divides the goals into detailed tasks for each group within the organization.

When goals and strategies overlap, the process requires setting up cross-func-tional teams. Initially, the efforts of these teams proved inadequate. "We found that employee-involvement teams didn't work well without training in brainstorming and problem-solving techniques," comments Kaspar Liepens, director of product quality assurance. To increase interaction, walls dividing various plant sections where differ-ent instruments were produced were knocked down, and the plant manager's office was located at the center of the manufacturing floor. Starting with one five-member team in 1986, the program grew to 24 ACE (achieving competitive excellence) teams with nearly 200 workers. ACE review boards, including Perkin-Elmer's president, meet quarterly to analyze progress.

Product design, one of the early targets for improvement, used concurrent engineering principles from the Juran Institute in Wilton, Connecticut. Perkin-Elmer adopted design-for-assembly software developed by Boothroyd and Dewhurst at the

University of Rhode Island for the design process. While going through all these procedures takes time, says Malandrakis, it reduces the time for the overall design cycle and forces the team to consider manufacturability, thus cutting parts counts (by some 50%) and engineering-change notices. The company also developed flexible-manufacturing methods to handle smaller lots more economically.

"Busloads of engineers went up to the University of Rhode Island for training," explains Malandrakis, "And they took some existing designs for classroom examples." The teamwork approach developed through this training led to the actual redesign of a number of the company's sophisticated laboratory analysis instruments. One engineer commented that the program stressed not only design for manufacturing and assembly (DFM and DFA), but also design that works (DTW) because of heavy emphasis on reliability.

Since honest self-assessment is vital in changing a corporate culture, receiving better feedback from vendors and customers is integral to Perkin-Elmer's quality drive. By working more closely with vendors, Malandrakis reports, the company was able to drop many operations—incoming inspection, for example—and cut delivery times for some products from 33 weeks to 8 weeks. Some poor vendor-rating scores turned out to be due not to failings by the suppliers but to problems with Perkin-Elmer's own blueprints.

Some components problems that had been blamed on vendors also were due to Perkin Elmer's own operations. This knowledge triggered programs to provide solutions. One of these programs, for example, is aimed at eliminating electrostatic discharge problems with some microcircuits. A video to show the proper handling of the devices was produced right in the factory and now is used to train the work force. Also, plant layout was adjusted so that final assembly was located much closer to test stations.

Cost accountants were assigned to the worker teams in order to allocate overhead more accurately. In building such instruments as gas chromatographs and spectrophotometers, direct labor accounts for a small portion (usually less than 15%) of the total cost. Capital equipment must be kept up to date to deal with fast-changing technologies. To boost competitiveness, notes Malandrakis, the company needs to find ways to invest and to change procedures to speed design, speed service to customers, and cut inventories.

Allocating costs is not an easy task. The company produces some 365 different instruments within 15 product lines; 1,100 units a month are made just in Norwalk. To better track costs and justify capital expenditures, Perkin-Elmer shifted to activity-based cost (ABC) systems. Under such an arrangement, chief financial officer Gordon Bitter works with manufacturing to show how to account for intangibles, such as product improvements or faster deliveries, in justifying capital expenditures.

One result was that employees convinced the company to buy an automated component-insertion machine. A conveyor also was set up to deliver special parts as they were needed under computer control. Optical alignments that used to take a day are now finished automatically in minutes, using new equipment that does not require hand tuning. After the company made such changes, production cycles were not only shorter but the number of finished instruments passing the final test the first time rose from 60% to 90%, even though intermediate inspection stations were eliminated to speed production, according to Malandrakis.

Perkin-Elmer also sought critical feedback through extensive customer surveys and took action to deal with major complaints. Among the top beefs were difficulties in ordering replacement parts and consumables and long delays in deliveries. As a result, the numbering system for parts was greatly simplified. Also, catalogs were made easier to use through the liberal use of photographs and blow-out, well-labeled diagrams. A new telephone system and added staff bolstered order-entry, so that an order can now get from the order department to the warehouse in 2 1/2 minutes. The company set up whole new business unit called PEXpress to serve customers and speed deliveries of parts and consumables. Any catalog order coming in before 3:00 P.M. can now be delivered the following day, and for rush orders the deadline is pushed to 4:30 P.M.

~ Zytec: Deming Methods Made a Big Difference in Quality

Since 1984, Zytec Corporation in Eden Prairie, Minnesota, has followed a total-quality business strategy based on W. Edwards Deming's famous 14 points (see the box). Zytec, a producer of custom power supplies, competes with dozens of companies, none of which owns a market share of even 10%. The company chose to use quality to differentiate itself from the competition. In Zytec's case, that was easier said than done. The company had been plagued by excessive scrap and rework, high inventories of both materials and work-in-process, and significant product-reliability problems. Its efforts to improve quality were fruitless until it adopted Deming's approach, says Ronald D. Schmidt, chairman and CEO.

The results were dramatic. Product quality, as measured by customers, rose from 94.2% when the new approach was instituted to 99.1% during the first quarter of 1990. Calculated mean time between failure improved from about 60,000 hours to about 300,000 hours. Revenue from new customers soared from $1.7 million in 1984 to more than $42 million in 1989.

Quality at Zytec improved so much that the company entered the 1990 Baldrige competition. It had to compete with major companies because its work force of 568 put it just over the 500-employee limit that defines a small business under the Baldrige rules. Although it was a strong contender, it did not rate a site visit. Still, the exercise provided priceless insights. "We learned our strengths and weaknesses," remarks John M. Steel, vice president of marketing and sales.

The application process itself consumed about 750 hours and cost an estimated $7,500 (aside from salaries), notes Schmidt, who managed the process himself. Along the way, Zytec executives picked up useful tips from other companies that had entered the competition. Motorola, for example, convinced Schmidt that a total-quality program of at least 5 years' duration was needed to prepare a strong entry. Xerox showed how a double-column format made it easier to get more meaningful charts into the 75-page entry form. Globe Metallurgical, the only previous small-business winner, explained the advantages of using a company data base to quickly retrieve key statistics, such as reduction in process variation. Once Zytec completed its entry, it set

∽ DEMING'S 14 POINTS FOR QUALITY

1. Constantly improve product and service to become competitive, provide jobs, and stay in business.

2. Adopt a new philosophy attuned to the new economic age. Western management must awaken to the challenge, learn its responsibilities, and provide leadership for change.

3. Cease dependence on inspection to achieve quality. Eliminate the need for mass inspection by building quality into the product in the first place.

4. End the practice of awarding business based on price. Move toward a single supplier for any one item, building a long-term relationship of loyalty and trust.

5. Improve constantly the system of production and service to improve quality and productivity and thus steadily decrease costs.

6. Institute training on the job.

7. Institute leadership to help people and machines do a better job. Leadership by management is in need of overhaul, as well as leadership of production workers.

8. Drive out fear, so that everyone may work effectively for the company.

9. Break down barriers between departments. People in research, design, sales, and production must work as a team to foresee production problems as well as potential problems in the use of a product or service.

10. Eliminate slogans, exhortations, and targets for the work force, asking for "zero defects" and new levels of productivity.

11. A. Eliminate work standards (quotas) on the factory floor, and substitute leadership.

    B. Eliminate management by objectives, numbers and numerical goals, and substitute leadership.

12. A. Remove barriers that rob hourly workers of their right to pride in workmanship. The responsibility of supervisors must be changed from sheer numbers to quality.

    B. Remove barriers that rob people in management and in engineering of their right to pride of workmanship. This means abolishment of the annual or the merit rating and of management by objectives or by the numbers.

13. Institute a vigorous program of education and self-improvement.

14. Put everybody in the company to work to accomplish the transformation.

Source: W. Edwards Deming

up a review process to learn how its performance could be improved even further with better measurements of progress. Feedback from the Baldrige examiners helped focus these efforts. The hard work paid off. In 1991, Zytec won the Baldrige Award.

The impact of Deming's principles is seen throughout the company. In the production area, an *Andon* board with red, yellow, and green lights is prominently displayed. A glowing red light means the line is stopped; a yellow light indicates a problem. Almost all orders show green lights, but when a red light does flash, the reason for the problem and the name of the person responsible are posted on a whiteboard. Problems range from major ones, such as "missing parts" for Zytec's just-in-time manufacturing to something as simple as "need new labels documented." While to workers in some environments the system could be embarrassing, Steel claims that it brings out the team spirit at Zytec. "People stop the worker in the hallway to ask how they can help," Steel says.

Apparently the company also made progress in removing the fear that can result in cover-ups or blame-shifting in a less tolerant setting. This is partly because of the move to cross-functional teams for each project. Also helpful is a stock option plan (for at least 1,000 shares) that extends to all employees, so everyone has a sense of ownership.

Another factor is a companywide training program. Every employee, from top executives on down, attends seminars on quality control, just-in-time manufacturing, and the ZIP, or "Zytec Involved People," program, which aims at moving the company toward participative management. Everyone learns at least the basics of statistical process control (8 hour course), but those directly involved in supervision or manufacturing take a more extended program (40 hours). Employees have input in setting up and revising training. In one survey, many employees requested more training in the use of some specialized test equipment. A new course was the result. Even workers' compensation is based on the number of tasks individuals are certified to perform.

The marketplace exerted tremendous pressure to speed design, says Steel, and Zytec trimmed down its design time from 5-6 years to 2-3 years. The push toward shorter cycles continues. Standard circuits were selected and fully characterized in computer-aided design (CAD) software. With standardized topologies and parts, first-pass design now is much faster, says Steel. Simulation is electronically linked to the CAD/CAM process, and the flow of design information was automated between the design center in Eden Prairie and the manufacturing plant in Redwood Falls, Minnesota, about 100 miles away. Electronic data interchange (EDI), now used to expedite purchase orders and invoices, is being extended to design review with customers.

Zytec feels one of its most significant accomplishments is a management by planning (MBP) concept based on the *hoshin kanri* practices of Japanese companies. The process marries the company's main long-range strategic plan with a yearly operating plan and with monthly, weekly, and even daily activities. Strategic issues devised by management are elaborated by six planning groups to produce 5-year objectives, corporate goals for departments, and a change-planning process involving all employees. Strategic objectives were for improvements in

- total quality commitment
- customer service
- profitability and financial stability
- housekeeping and safety
- employee involvement
- shortened product cycles

Schmidt now feels the company is well on its way toward semiautonomous, self-managed work groups. This is in sharp contrast to the situation in 1984, when there was heavy dependence on inspection, management by objectives, and large safety stocks. He terms that condition "unconscious incompetence."

First, he says, it was necessary to progress to "conscious incompetence," in which shortcomings were recognized as the entire group studied such methods as JIT, Deming's principles, and ways to eliminate waste. Zytec now is reaching "conscious competence," with broad emphasis on continuous improvement, reduction of cycle times and inventories, training of multifunctional teams, improved yields, and management by planning.

The ultimate objective is what Schmidt terms "unconscious competence," based on a total culture change that leads to automatic problem solving by all employees, self-managed work groups, and MBP as an accepted way of corporate life.

Chapter 11

Electronics Manufacturers Strive for Quality

Most U.S. electronics manufacturers face dual pressures for world-class quality. First, they must compete effectively in the global arena, often facing off against either subsidiaries or suppliers to the Japanese powerhouses in computers and consumer electronics. Second, their major U.S. customers have become much more quality conscious of late, looking to their suppliers to strive to meet the demanding criteria set for the Baldrige Award program, if not to actually enter the competition.

At hundreds of electronics plants all over the nation, even a casual observer will find that not just management, but workers and supervisors as well, are all pulling together to meet the challenge. The following reports are based on visits to some of these plants.

At Hewlett-Packard (HP) in Silicon Valley the process starts right at the top, with the way upper management goes about setting and communicating overall strategy, and moves on down through the ranks. Across the country, nestled in Pennsylvania's picturesque Amish country, AMP, Inc., a leading maker of connectors, conducts its own quest for excellence, ranging from the process for designing new products to making sure phones get answered quickly and orders get shipped on time. Back in Silicon Valley, Conner Peripherals meets and exceeds ever-tougher quality goals in spite of record-setting growth. Even in the out-of-the-way hills of Logan, Utah, workers at Bournes put signs on the walls urging co-workers to do a careful job of tracking process variables. Here are their stories.

❦ *Hewlett-Packard: Doing It Right the First Time Saves Millions*

It took a turnaround story for Hewlett-Packard Co. to finally implement a companywide quality program. In 1977, Yokogawa Hewlett-Packard, HP's instruments and computers joint venture in Japan, was losing market share to higher quality Japanese competitors. In fact, even when benchmarked against other Hewlett-Packard units around the globe, Yokogawa-HP had the lowest quality scores.

Through conscientious application of the quality principles of Genichi Taguchi, a leading international quality disciple, and others, Yokogawa turned its quality around in 5 years to win the prestigious Deming Prize in 1982.

These lessons were not lost on the company's Palo Alto, California, headquarters. In 1983, HP appointed Craig Walter corporate quality director, and soon the quality approach was introduced to all of HP's worldwide operations.

Along with other benefits, implementing quality throughout the company yielded an estimated $400 million in savings of warranty costs, due to a renewed emphasis on hardware quality, Walter says. A beefed-up emphasis on design for manufacturability resulted in another $400 million in savings, Walter estimates. This was all achieved because HP president and CEO John Young wanted a tenfold improvement in hardware reliability during the 1980s. Thanks to a rigorous quality program, he's getting it.

After the edict from the top, progress in improving hardware quality was rapid in the 1983-86 period. "When you first get started, it is easy to improve," says Noriaki Kano, a Japanese professor who began to work with HP's Computer Division in 1985. "You are mining the surface gold."

Defect levels that were 900 parts per million in 1983 had dropped to the 100 ppm level by 1987-88, according to Carl Kolich, components group quality manager. But then they began to level off, and now HP is having to reach even deeper to gain the final increments of quality improvement it seeks.

At the same time, the company is moving well beyond product quality in seeking areas for improvement. Using quality function deployment, statistical process control, and an increased emphasis on customer satisfaction, HP management has moved quality principles through every level of the company.

The quality push is paying off. As an exercise early in his tenure with HP, Walter had some divisions estimate what their costs would be if there were no defects in parts or processes, deliveries were all on time and trouble-free and there were no difficulties in using HP products in any application. He found that these nonquality costs added up to 25 to 30% of sales dollars. On top of that were design inefficiencies, scrap, and rework that, if eliminated, would reduce the average time-to-market by a third.

The results of these studies indicated that the company could finance growth of 50% a year (assuming products gained market acceptance) and greatly improve profitability just from savings on process improvements. Since that time, Walter estimates, HP has reduced its inventory levels by $500 million, accounts receivable by $150 million, and time-to-market by several months. Yet he believes that the costs of not achieving total quality remain at the 25 to 30% level. That means these costs are much higher today than when the survey was done because the total business is so much larger now. He suggests, for example, that 1988 distribution costs of about $1 billion could be cut in half with improved methods.

Under Walter's direction, the quality process was unified across all of HP's units, using a consistent set of tables for planning, setting up goals and measuring performance, defining tactics for implementation, and reviewing progress toward objectives. Abnormality reports detail how deviations from targets are handled. The quality reviews are now the basis for a mini-Malcolm Baldrige National Quality Award competition among HP's units. "There's no big emphasis on the winner," says Kolich, "but nobody wants to come in last!"

"Quality is the one thing left out of the model," adds Kolich, who is also quality manager for HP's Optical Components Division. "We want it to become so much a part of infrastructure that we don't even need quality specialists. We are trying to work our way out of a job!"

"The emphasis has shifted instead toward Joseph M. Juran's definition of quality: 'Quality equals fitness of use by the customer,'" explains Kolich. For many years American industry equated quality with conformance to specifications, he explains, and with the rapidly changing expectations of the marketplace, "specsmanship" is no longer acceptable.

Product development teams now include a range of specialists in addition to design engineers, including representatives from manufacturing, marketing, and, if the product is to be sold in Southeast Asia, for example, a representative from that area. Hewlett-Packard now shares much more information with its key customers and is developing similar partnerships with some suppliers. A few trusted vendors now ship directly to the production line for just-in-time manufacturing, according to Kolich.

∿ Adjusting Corporate Strategies for Continuous Improvement

Quality function deployment, a process aimed at structuring HP's operations to best meet customer needs, is now widely used in the company. The Yokogawa-HP quality process included what the Japanese call *hoshin kanri*, a methodology for policy deployment that can achieve quantum leaps over and above the continuous improvements typically associated with Japanese industry.

Hewlett-Packard is now adopting the *hoshin* approach, which was introduced to Japan by quality pioneer W. Edwards Deming in the early 1950s. Deming's view was that making people work harder, or watching out for defects more closely, would do little to improve quality. Instead, he counseled, the entire system must be reshaped to deal with root causes of problems. Using the *hoshin* approach, HP periodically reevaluates and modifies strategies to achieve continuous improvements as well as set new priorities. In brief, the process goes through the following sequence:

- *Plan.* Articulate detailed plans based on objectives, goals, and measures.
- *Do.* Carry out the plan for a specified period (which may cover 3 to 7 years).
- *Check.* Conduct formal reviews, check progress toward goals using measures, explain deviations (positive and negative).
- *Act.* Take corrective action, which may, in turn, lead to a new cycle beginning with revised detailed plans.

Such procedures are now part of HP's annual quality reviews. The company redefines each business unit's goals each time the cycle is repeated. It bases revisions on a wide range of inputs, including overall corporate objectives and external factors, such as economic conditions and competitive forces, as well as on customer feedback and surveys. One of these is a worldwide customer satisfaction survey, first launched by HP in March 1988, when questionnaires were sent to 40,000 customers all over the globe.

Each approach to quality improvement introduced into the company must be honed to achieve desired results, points out Maarten Kalisvaart, quality manager for the Optoelectronics Division. He emphasizes that there is a wide difference between statistical quality control (SQC), the method used in the past, and statistical process control (SPC), which HP has now adopted. "Statistical quality control was like measuring how muddy the river is," he explains. "By contrast, statistical process control is a way to move upstream to find out where the mud is coming from."

Getting engineers or operators to respond properly to the control charts used to monitor processes is a problem, he says. If everything is within desired limits, they should not make adjustments. But if the charts indicate problems are developing, Kalisvaart wants the operators to write notes on the charts, speculating on what the causes might be and stating what actions will be attempted to correct the problem. "I like *messy* control charts!" stresses Kalisvaart.

Kalisvaart agrees that Taguchi's designed experiments are a valid way to find out the effects of various manufacturing process variables without lengthy, exhaustive testing, but he believes that once these experiments have been done, more conventional Western mathematical techniques are preferable to Taguchi's methods for analysis. At the same time, he credits Taguchi with having great influence in improving production processes all over the world.

Software is one of the biggest areas for potential improvements. About 70% of HP's research and development staff works on firmware and software, and 40% of those are involved in fixing or enhancing software (often a euphemism for overcoming a weakness in the original design), according to Walter.

Eliminating the estimated 25% of engineering effort now devoted to such rework activities would free 750 engineers to work on new products, he points out. Working with the Yokogawa-HP group as a model has also shown the company that the Japanese are much more disciplined in software development, with very low defect densities in writing code, better documentation, and an organized approach to reusing code for subroutines and other common features.

∾ Software Goal: Ten Times Better in 5 Years

Because so many of HP's engineers, including some in every division, now work on software, points out T. Michael Ward, manager of corporate software quality, it's not surprising that CEO Young has now set a new goal of a tenfold performance improvement in HP's software quality over the next 5 years.

"Many organizations back into being software companies," Ward says. The Jet Propulsion Laboratory, in Pasadena, California, for example, now has five software engineers for every hardware engineer. HP recognizes that software has become an integral part of almost all new developments in the company. Because more than half of HP's annual sales are now derived from products developed over the preceding 3 years, it's no wonder software has been chosen as the chief executive's next push, according to Ward.

Part of the quality effort is a more systematic approach to developing and documenting new software, based partially on what the company learned by working with Yokogawa-HP. Another is a goal to achieve a 12% improvement each year in the rating for software in the customer satisfaction survey. To help achieve this, HP has put together a software program called *FURPS+*, which encompasses a wide range of additional quality efforts, according to Ward.

The FURPS part of the program includes efforts to improve functionality, usability, reliability, performance, and supportability. The plus includes such elements as predictability (how well products can be engineered to meet specific goals), portability (ability to run programs on a wide range of hardware platforms), and localizability (ease of converting software to foreign languages).

As with other HP programs, FURPS has a strong focus on customer needs, says Ward. HP developed a new pseudo-language, for example, that makes it easy to construct icons representing processes, such as liquid levels, transducer outputs, or pump operations, in easily interpreted graphic form. Another goal is to make it easier for what Ward calls the "middleman" to tailor software products for specific applications within a user organization.

The willingness of American firms, even competitors, to exchange ideas on quality efforts such as this has increased greatly since about 1986, according to Walter. Hewlett-Packard, for example, has stepped up the benchmarking of its products versus the "best in the world," and has been learning from Baldrige Award competitor Xerox Corporation, which has been knee-deep in the quality fray since 1980.

Walter sees the Baldrige Award competition as a major step toward improving corporate cooperation, so that U.S. firms can help each other become even stronger competitors in the global marketplace. HP's Roseville Terminals Division entered the competition in 1989, but the company chose not to enter in 1990. The internal mini-Baldrige competition will keep Walter up to date on progress throughout the company so that he can make a decision about entering in the future.

∿ AMP Brings Quality to the Countryside

The world's leading connector maker employs a strong work ethic and a streamlined organization to boost quality management and cut costs. Nestled in the pastoral capital city of Harrisburg, AMP, Inc., the leader in the worldwide connector market, has been likened to a Japanese company in the middle of Pennsylvania. Its quality-oriented culture is not just a fad but stems back to the firm's earliest days. After establishing AMP in 1941 in Elizabeth, New Jersey, founder U. A. Whittaker moved the headquarters close to the Lancaster Valley to tap the strong work ethic and high integrity of the Amish labor pool. Today, a formal quality improvement program begun in 1983 has permeated operations from manufacturing to customer service. Through improved controls, the $2.7 billion company estimates it saved $60 million in quality costs over a recent 5-year period.

"Cost reduction extends from the factory floor to white collar management and ties into our quality program," says Harold McInnes, AMP's vice chairman. AMP contained commodity item costs by buying globally and by reducing its supplier base by about 20% over 3 years while building closer relationships with the chosen vendors, he adds.

Cutting costs was a critical challenge for AMP, which strives to maintain quality while at the same time sustaining a 15% compound annual growth rate, with 18 to 20% pretax margins. Between 1985 and 1986, AMP shrunk its domestic work force by

3,000, downsized operations and consolidated into larger plants, and streamlined its catalog of products by the tens of thousands.

One of the many changes at AMP was a renewed emphasis on quality. Since 1984, the company has tracked quality through dozens of performance measurements for both domestic and international operations. Those gauges measure such areas as scrap rates, defective parts per million, rework, shipping errors, returned goods, customer complaints, and order handling times.

AMP's goal is a quality level of 100 defects or fewer per one million parts. The company claims that with its improvements in quality, AMP customers accepting products on a ship-to-stock basis increased from 9 to 32 between 1987 and 1988.

～ "Plan for Excellence" Leads Quality Push

To reach its goal, AMP recently adopted several new approaches to quality management. It entered the Malcolm Baldrige National Quality Award competition in 1988 and used the award guidelines as the basis for its strategy to compete again in 1991. It created a managerial position, corporate vice president of quality assurance, to work closely with a newly appointed "plan for excellence" steering committee of top-level executives, chaired by McInnes.

Other recent developments in quality management are

- an engineering excellence program that sets standards and acts as a quality checklist to encourage superior product design
- an engineering education program of 26 courses offered to 1,500 engineers, scientists, and technical supervisors
- a value added manufacturing program teaching just-in-time methods, total quality concepts, and total employee involvement techniques to 10,000 employees
- a corporate logistics program to improve product delivery performance through newly centralized warehousing, better inventory control and delivery performance measurement.

According to president Jim Marley, AMP raised its finished goods inventory turns by 20% over 2 years and has pushed for better customer delivery service at a lower inventory cost. A flagship of AMP's quality program is its "scorecard" program for on-time delivery. AMP defines on-time as shipping up to 3 days early and no later than the promised delivery day. Overall delivery performance went from 65% on time to nearly 90% in less than 2 years. For military products, AMP strives for 4- to 6-week lead times, down from 20 to 22 weeks.

AMP's corporate logistics department was restructured in 1987. "We're trying to differentiate ourselves by quality," says J. Keith Drysdale, director for corporate logistics. "The whole idea is to manage products from the cradle to death. You get a much better flow and support of goods." That logistics effort, he estimates, costs about 15% of sales.

Performance is now measured in foots, or factory orders on time. AMP guarantees delivery dates of less than 2 days for about 8,000 parts. To meet that pressure, the company automated its call distribution system at a cost of $1 million. Within 60 seconds, calls are rerouted; calls that overflow from customer service operators in Harrisburg automatically ring in AMP offices in Chicago or Valley Forge, Pennsylvania. Previously, "it used to be very often callers got busy signals," or calls were abandoned, admits Drysdale. He estimates the system increased productivity in processing line items by 25% per operator.

The logistics department also centralized warehousing and professionalized warehousing procedures. Before, AMP operated many small warehouses tucked in the back of its factories. Because of AMP's scorecard program, Drysdale knew deliveries were running 20% behind schedule, but he didn't know why. "It was like having a 103-degree temperature and not knowing what the disease was," he remarks. A revamped tracking system showed that mislabeling was leading to a high volume of customer returns. Professionally managed, "the returns as a percentage of sales has fallen by a factor of ten," according to Drysdale.

Quality isn't just a concern of senior management. Over the past 5 years, AMP's domestic employees submitted more than 11,000 quality recommendations; when acted upon, some 85% led to improved quality, productivity, and customer service. The framework for improvements is a network of 130 quality improvement teams, organized by function, that are responsible for setting goals and correcting problems. AMP will need that team effort as it tries to increase sales of its high-performance value-added products. For the buyers of these connectors, quality is more important than price.

To succeed in the "smart" connector market, customer service and close customer ties will be increasingly important as AMP's bread and butter high-volume lines face steady price erosion. Cost-cutting measures can only go so far in keeping margins up as prices drop. Better quality just might be the ticket to help the company shift more business to higher-priced custom connectors—and to maintaining AMP's 15% annual growth, according to vice chairman McInnes.

∿ Conner Peripherals: Fast Drive to Quality

Talk about taking off like a rocket! Here's the scorecard for Conner Peripherals, Inc. of San Jose, California: the first full year's sales broke the record for U.S. industry, hitting $113 million in 1987. Then sales more than doubled in 1988, to $257 million, followed by 1989 sales that surged more than 2.5 times more to hit close to $700 million!

Since Compaq Computer Corporation of Dallas, the previous record holder for rocketing into the elite ranks of the Fortune 500, only hit $625 million in its third year, Conner has become the newest maxi-success story of the U.S. electronics industry.

The secret? Where many companies are still shifting from technology-driven to market-driven strategies to keep up with rapid change and global competition, a few firms are shifting even beyond, to customer-driven strategies. By working hand-in-glove with key customers these innovating companies are getting out ahead of market changes and helping to create the future rather than waiting for markets to develop and then hopping aboard.

Nowhere is this strategy more apparent than at Conner Peripherals, the world leader in the pipsqueak-sized Winchester hard-disk drives (2 1/2 and 3 1/2 inch) built into laptop and portable computers, the fast-track segments of the otherwise sluggish microcomputer market. Conner calls its approach the "sell, design, build" strategy, based on forming alliances with original equipment manufacturer (OEM) customers for joint product designs. The company's stock-in-trade is getting out front in the drive market, particularly in the tough-to-build, low-power, high-capacity palm-sized hard drives that make bantamweight computers so appealing to buyers. In fact, helping customers into tomorrow's markets has been a way of life for company founder Finis Conner, who helped Allan Shugart start Shugart Associates and Seagate Technology, both drive makers.

In the mid-80s, Conner got a cool reception when he went after venture backing the third time around, this time joining with John Squires, another entrepreneur who had earlier founded and left Miniscribe, to produce an innovative 3 1/2-inch drive. Profits were tough for disk-drive makers in those days as new companies proliferated and many failed, scaring those who had bankrolled previous start-ups.

Rod Canion, an old friend of Conner's and co-founder and CEO of Compaq, needed a mini-disk drive for a new computer his company was planning. The two met at Comdex, a major computer trade show, in the spring of 1986, and out of their chat came a joint design project and $12 million in backing for Conner. Compaq bought 90% of the new disk-drive maker's first year's output and still owns 40% of Conner.

After a bang-up start, Conner faced a new problem. It had to convince other makers of laptops, portables, and desktop micros, some of them competitors to Compaq, that Conner was a legitimate new player in the tough disk-drive market and not just a captive supplier for its primary customer in Dallas. A major part of Conner's solution was a push toward being tops in quality as well as growth.

How does one instill a quality culture in a company that is in the process of growing from zero to more than 4,400 people in 3 years and is rapidly putting new plants in San Jose, Singapore, Malaysia, and Italy (for European markets) to go with its research facility in Longmont, Colorado? Even with a quality team in each plant, management recognized that it was essential to develop a uniform quality management system throughout the company. So Simon Ancri, a 25-year veteran of IBM, came in to head a small corporate quality staff at the San Jose headquarters as vice president, corporate quality and reliability.

The problems Ancri faced were ones that will be increasingly common in the fast-paced electronics business. "I like the market extremely fast," boasts Finis Conner, now chairman and CEO, "While my competitor is trying to figure out where to go next, I'm running away from him." Achieving enough agility to keep that rabbit out ahead of the hounds challenges every sector of the corporation, and because quality problems tend to be much stiffer for pathfinding products, it makes the quality guru's job doubly tough.

To increase flexibility Conner assembles its drives mostly from subcontracted components and subsystems. That means close coordination not only with major OEM customers, where Conner often participates in new product designs, but also with a wide array of suppliers. In the early days there were too many suppliers, Conner admits, so the number was culled, and the firm now has a subassembly plant of its own in Penang, Malaysia.

Conner's three-tiered approach to rapidly ramping up new products to high-volume production is another factor that really keeps Ancri's quality team hopping. The research and development (R&D) team under John Squires, executive vice president R&D, in Colorado, puts together prototypes Then pilot plant production moves to the company's plant in north San Jose, where the company sets up such start-up details as debugging, documentation, and testing procedures. (The company hoped eventually to be able to shorten time-to-market even more by eliminating this phase.) Finally, when production ramps up, the action moves to Singapore, where output ranges up to 10,000 units a day. All three stages of the prototype-pilot plant-volume production sequence have to bop along concurrently as the company moves from drives for lightweight desk-tops to portables to lap-tops down to notebook-sized computers. To top this off, Conner also has developed pint-sized, 1 1/2 inch high, 210 megabyte-plus 3.5-inch drives for workstations and advanced desktops. "We're 3 years old and we're already moving into our fourth-generation products!" says Ancri.

Even though there are quality managers at each plant, any feedback from customers comes through Ancri and his corporate group, and they are responsible for policies and procedures that ensure that quality problems are quickly analyzed and action plans developed and carried out and then followed up to ensure that the cure has worked.

Product quality starts with Conner's basic design strategy, according to Ancri. Designers embed disk controllers in the drives, halving power requirements, and they use customized microprogramming to replace the tangle of mechanical parts and added electronics needed in bulkier disc drive designs. One reason for the innovative design philosophy, Ancri says, is that most disk drive designers are electrical engineers or mechanical engineers, while Squires is a physicist with less committment to any particular engineering discipline.

Fewer moving parts and chips cut down on potential quality problems, and Conner's engineers work closely with product and manufacturing engineers at OEM customers to tailor their drives to manufacturing and system requirements. Design for quality is only the start, however, Ancri recognizes.

Thorough testing at every stage of the manufacturing process goes one step beyond many programs, for example. Even after drives have been packed up for shipping, Conner does what Ancri calls an "out-of-box audit." Sample drives are taken out of the packaging for a test simulating a customer's incoming inspection, thus determining if the packaging process itself may be causing a problem.

If workers detect trouble at any stage of production, they have the right to shut the line down until corrective action has been taken. Sometimes, however, problems are not detected until the drive reaches a customer. To ensure rapid feedback to any such problems, and to collect complete data on any failures that do occur, Conner assigns quality engineers to major customers. There is also almost daily communication between Conner and these customer sites, according to Ancri. It goes both ways, because Conner freely shares its own process and test data with customers, as well as the results of its investigations of any failed drives that have been returned for analysis. To speed this data-sharing process, Ancri is now investigating setting up compatible facilities so data can be transmitted instantly between Conner and key customer's plants.

The manufacturing team involved in any area where a problem is detected uses process data to investigate root causes, develops a corrective action plan, and then sets its own tough targets for improvement, Ancri explains. Although every effort is made to reach these stringent goals, Ancri says that in this phase of building a quality culture he is much more concerned with building the right attitude throughout the organization than with the specific numbers. Still, the numbers are carefully tracked, with results posted in work areas. For a quality culture to pervade an organization, the attitude must start at the top. "There must be consistency of purpose from the executive level on down," says Ancri.

The goal is continuous improvement toward greater customer satisfaction. Crucial to making customers happy, Ancri recognizes, is quick response to any perceived problem. Top management has added impetus to Ancri's drive for faster response with a system set up by William J. Almon, another long-time IBM veteran who joined Conner as president in early 1989. Almon had boxes with a red light installed in the offices of key executives, and when customers report a problem the light keeps flashing until the problem is resolved.

Even with all this diligence, sometimes tracking down problems to their source can't be rushed. Ancri cites one case where a customer kept reporting trouble in drives that had checked out fine at Conner. Days of observation by the quality engineer at the customer's facility finally uncovered a slight difference in the way the customer performed tests. Conner immediately changed its test methods to match those of the customer, and Ancri believes that the root problem was therefore found and corrected. Careful tracking continues to make sure.

Cross training enhances the development of partnerships with customers and suppliers; Conner sends its people to training sessions with key customers and suppliers and opens its own training programs to customers and suppliers as well. Communications on quality take place almost daily with all Conner's scattered plants as well as with customers, and Ancri says he is now working to extend the system to key suppliers as well.

On a monthly basis, Connor performs careful analysis of the cost of quality—rework, scrap, repairs, breakdowns, and so on. Ancri tracks the cost of quality by product type, by location, and by type of failure and identifies who might take action to reduce the costs in each case. His team has set aggressive targets for cost-of-quality reductions. "We need to make the data on failures highly visible, and then increase the accuracy of our cause analysis," Ancri explains. "It's a layered process, like peeling an onion."

Even with the constant communication, Ancri himself visits the Singapore plant every 5 weeks or so to check on quality progress. An hour before leaving on a recent Saturday morning, he learned of a problem from one of Conner's customer reps. He called the managing director of the Singapore plant before taking off, and the quality manager there immediately began an investigation. Ancri says that by Monday morning the quality manager had faxed a preliminary report on the investigation to San Jose.

The continuous drive for better quality cut failures by three times for some products over the last half of 1989, according to Ancri, and goals were set for 5 to 6 times additional reductions by the end of 1991. "You need visible goals, and then measurements that show how well you are progressing toward them," says Ancri. Efforts to improve are not always successful, so everyone must be prepared to readjust when necessary. When you reach the goals, you have to set new ones.

"You have to keep raising the bar," says Ancri, because the competition is always breathing down your neck. That's particularly true in the disk drive business. Not only have U.S. manufacturers steadily boosted storage capacity and reliability while lowering costs, but tough Japanese competition, particularly from Sony, is now entering the market.

Still, Conner takes pride that not only does it sell to American computer makers like DEC, Apple, and Sun, and European ones like Olivetti, NCR Europe, and Zenith, it also has brisk sales to Japanese firms like NEC, Sharp, and Toshiba.

∽ Bourns: Acceptable Isn't Good Enough

At a small factory in Logan, Utah, some 85 miles north of Salt Lake City, the Bournes Networks, Inc. (BNI) staff has a new slogan that's really caught its fancy: "Without the data, you're just another clown with an opinion." The data referred to are statistical process control (SPC) figures manifest on record sheets that festoon the factory walls and the busy clipboards of employees who by now probably log data in their sleep.

The obsession with SPC data is not wasted. The BNI division in Logan—which makes mostly thick-film molded and conformal resistor networks and resistor-capacitor networks for its parent, passive-components giant Bourns, Inc.—has, since 1985

- trimmed average defect rates from 162 to 64 parts per million
- showed double-digit growth rates in an industry where 9% yearly revenue gains look great
- doubled market share to almost 25% of the U.S. resistor-networks business
- helped drive new product development: 45% of sales come from products designed in the past 3 years
- cut product introduction times to gain new market opportunities.

∽ BNI Aims to Enter the Baldrige Competition

The all-around improvements at BNI and at a sister division have convinced top management to enter Bourns in the Malcolm Baldrige National Quality Award competition in 1992. In the early 1980s, Bourns, like many electronics manufacturers, trumpeted quality as a basic tenet of doing business without always following through. For many, the early-80s view of quality meant avoiding customer ire by ensuring that most products worked when they arrived and by groveling to take back the units that had arrived defunct. The *lingua franca* of quality was AQL—acceptable quality levels.

Bourns Networks, which today has more than 400 employees and annual revenue nearing $50 million, wasn't that different—and it showed. "We were really a nonentity in the market, to be honest," says Bing Harding, who joined the then 6-year-old operation as its president in 1982. Harding inherited a factory where payroll and related costs for hourly workers made up a dizzying 62% of sales and where quality had taken a back seat to order filling. "There was no way we could survive," he says.

Survival was not what Bourns had in mind. The new strategy was for BNI to expand its networks business dramatically, and the chosen tactical tool was a push on quality. "We've set a corporatewide objective of improving quality by a factor of 10 by 1994," says BNI's chief executive Gordon Bourns.

To fulfill the ambitious new mission, the company made several congruous decisions. "Acceptable quality" procedures would be succeeded as quickly as possible by an emphasis on parts-per-million defect levels and then by a thorough statistical process control strategy. The company would make heavy investments in automated manufacturing equipment. Finally, Logan would grow from a pure manufacturing site to become BNI's business-unit headquarters, encompassing design through marketing, with a facility in Cork, Ireland, as a satellite manufacturing site. (Previously, BNI had been spread out nationally across 13 buildings in various sites.)

An initial step was to tighten existing acceptable quality levels; 1982 saw end-of-line defect levels of 2.5% on physical parameters and 1% on electrical. Four years later, the defect rates on physical parameters—pin lengths, pin cross-sections, coating integrity, and so on—had dropped to 0.65%, with electrical defects in only one out of a thousand parts coming off the line. By 1984, BNI had shifted emphasis from inspecting out problems to having fewer problems to inspect. The result was the redeployment of all gate inspectors. For the sales years of 1983 and 1984, the company was able to breathe a little easier. It was not to last; Harding remembers watching the post-PC market evaporate in 1985 and the meetings during which company managers fretted, as prices and volumes shriveled, over the demise of profitability.

While Harding was lucky to have access to patient money in the shape of a privately owned parent, Bourns's long-term focus was not the only performance lubricant. "The networks unit has definitely benefited from the poor performance of its competitors," notes Ralph Anavy, president of Electronic Outlook Corporation, a research firm that tracks the passive-components business.

While that may be true of some that may have suffered rather than benefited from acquisition, it doesn't apply to rival CTS Corporation of Elkhart, Indiana, where similar quality programs give Harding and his staff a run for their money. The motivation is no different from at Bourns: quality has become a customer requirement and no longer simply an advantage.

Betting on the fast-growth prospects of the networks sector—*Electronic Business* magazine puts it at nearly 10% annually over the next few years—Bourns opted to centralize all networks activities in Logan. Harding's appointment as the top executive there led quickly to spending, from 1982 to 1985, of about $6.5 million on production equipment to reduce labor content and lay the groundwork for the statistical process control program that was to follow.

The statistics offensive began in 1984 and 1985 with what Harding calls a lot of definition—of product, of process, of operating procedures, of training, of customer liaison. Formal implementation kicked off with indoctrination in statistical techniques and in the 14-point gospel according to quality messiah W. Edwards Deming: 12 hours of it for hourly paid workers, 40 hours for exempt staff.

With formal education complete by late 1987 (training cost BNI more than $500,000), Bournes set up improvement project teams in engineering, manufacturing, quality, and customer service, the objective being to identify unstable, or highly variable, processes, pinpoint causes, and come up with solutions. The teams began to post SPC charts that tracked such process variables as incoming material quality and machine calibration.

With process stability understood, employees turned to the tough bit: getting the processes to conform to new statistical limits, and to meet narrower and narrower variability. Throughout, the key was in having every employee understand what factors—a laser-scribe depth, a dirty resist-screen mesh, a molding temperature—have what effects. Putting a tourniquet on process variability had striking results. BNI accomplished its 1988 goal of 100 parts per million defects, down from 162 ppm in 1985 when production capacity was half 1989's levels. The 1990 goal of 50 ppm was exceeded before the end of 1989.

Customers appreciate the pickup. "Once we sell a Bourns part, it stays sold," notes Michael Morton, director of passives marketing at TTI Inc., a components distributor. Within a year, one telecommunications customer expects not to have to inspect any incoming Bourns networks as it moves BNI onto its audit schedule.

Market surveys credit Bourns and its competitor CTS with nearly 25% each of the $220 million-a-year U.S. resistor-networks market; in Bourns's case, that is up from about 12% in 1985.

A walk around the Logan facility unearths clues that statistical process control has reached the work force. Preaches one March memo from a line worker: "To all thick-film operators: SPC is a way of life! Live life to the fullest—fill out your charts!"

The operators have launched initiatives of their own: one group recently gathered data to beat a problem with irregular cross sections on network pins. The effort pinpointed tool changes as the path to improvement, so more frequent tool swaps were accommodated.

The quality gains have, according to BNI, helped the unit regain solid profits. They have also opened up new business opportunities by allowing faster deliveries of new-design engineering samples to beta sites, which makes it easier to become qualified.

Hitting shrinking market windows with dependable product is vital to customers whose products quickly age, such as makers of 3 1/2-inch hard disk drives, graphics hardware, and minisupercomputers. For BNI, the payoffs are real. One personal computer customer is likely to follow $2 million in business with another $3 million order, and the prospect of a further $700,000 for new products.

The challenge now is to push further than the 10 ppm overall goal for 1992. A certain amount of fear is no bad thing. "We have to assume we're not in the lead on quality if we're to keep being good at it," says Harding. He is right to worry: archrival CTS claims to be shipping some parts at 20 ppm already.

Chapter 12

Small Electronics Firms Join the Baldrige Quest

The battle for global markets is nowhere tougher than in the electronics business, especially for the smaller firms that supply parts and subassemblies to the major companies. The U.S. electronics industry faces a daunting challenge from both Asia and the European continent.

Japan's powerful electronics giants are vertically integrated, with the systems-level divisions that build computers, office equipment, robotics, and consumer gear pressing their semiconductor groups for advanced technology and lower production costs. Most of these powerful international giants are also members of multifaceted families of companies, called keiretsus, which include banks and insurance firms. These financial groups can provide the manufacturing arms with low-cost capital. In electronics, Japanese companies have gone far beyond the copycat label once pinned on them. National research initiatives coordinated by the Ministry of International Trade and Industry (MITI) and other agencies help push the state-of-the-art in key commercial technologies. Japanese business customs, along with MITI policies, have discouraged manufacturers from buying imported products whenever there is a domestic substitute, making it tough for U.S. firms to penetrate the huge market in Japan.

In Europe, big electronics companies have consolidated, and major joint research projects are conducted across national boundaries. Further moves toward European unity in 1992 are likely to increase trans-European collaborative activity and make it tougher for outside firms to gain or hold market share.

American electronics firms, undisputed global leaders just a couple of decades back, have been struggling to meet this tough competition, but many observers point out that U.S. government policies toward R&D and trade have been driven essentially by military and geopolitical interests rather than the commercial marketplace. Leaders in the electronics industry see the push for quality as perhaps the last best hope for American companies to retain, or to regain, competitive strength in global markets.

For the American electronics giants to achieve world-class quality it isn't enough for them to develop an internal culture driven by total quality management processes. They also need suppliers that are just as diligent in applying the principles of total quality management. Japan has a big head start, with more than 4 decades of experience in practicing and even extending the principles of statistical process control and continuous improvement espoused by Deming and other gurus, not just in their own operations but through the ranks of their supplier-partners. That's one reason, aside from cost advantages, that U.S. firms buy so many parts and subassemblies from Japanese sources.

Now U.S. vendors are fighting back. Smaller American electronics firms remain vibrant and innovative, but they are rapidly gaining strength in total quality management as well. Following are the views of a number of small-company executives on the push for quality in this market segment. A closer look at the efforts of a few individual firms will show how even organizations with limited resources can develop programs aimed at achieving world-class quality. Sputtered Films and Vamistor are among those that have set their sights on winning a Baldrige Award. A quick look at what other small electronics companies around the country are doing to adopt total quality principles helps illustrate the breadth of the movement now sweeping American industry from top to bottom.

∿ Electronics Industry: Small Firms Simplify Quality

For small electronics companies, quality is not an option. "If we didn't have quality, in products and service, we would go out of business," Robert Swanson, president of analog chip maker Linear Technology Corporation, says bluntly. "It's that simple."

What is an option today is how small companies put quality programs in place. Whether they hire consultants (sometimes at a whopping $2,500 to $10,000 per day), attend seminars or "quality colleges," or develop their own programs, small companies are finding creative and ingenious ways to match or outshine their larger competitors.

Quality programs too often have a tendency to become overcomplicated, complains Robert Graham, president and chief executive of chemical vapor deposition company Novellus Systems Inc. "Pretty soon your major goal is paperwork. That's a lot of formality in a small company." Graham, who established zero-defects principles in his company, speaks for a number of executives who find the quantitative side of quality programs a bit much for a small company.

The key is simply getting employees to realize that if something is wrong, it needs to be fixed so it doesn't come back to haunt them later, says Tom Rohrs, manufacturing vice president at Mips Computer Systems, Inc. "Quality becomes everyone's job."

Executives say starting a quality program can actually be easier for a smaller firm, thanks to the limited number of players and the relatively close communication. "Small companies can put quality in early on," says Visix Software Inc.'s chairman Jay Wettlaufer. "You can't go back and put quality in."

Communication is so easy at some companies that they can use consensus management to implement quality decisions. At Exabyte, a 1987 start-up which manufactures tape subsystems, "We use consensus opinion to solve problems," quality assurance manager Chris Wening says. "Our psychology is that everyone should be

↝ Some Tips for Putting Quality on the Road Map to the Top

For small companies still undecided about how to pursue quality, there are a few basic rules to keep in mind, according to John Wood, president of the management training and quality consulting firm Quality Now in Glen Ellen, California. Small companies always think they are different from large firms, so they need different programs. Wrong, Wood declares. "Once the founders hire their first employee, they are on their way to looking just like a large company," he says. So the first rule is to learn from larger companies. Here are some tips on how to proceed:

- Talk to someone who has instituted a quality program before, either a consultant or someone from a large company who has been through it.
- Make sure the top person in the company is 100% committed to quality.
- Don't confine the quality program to manufacturing.
- Understand the internal quality situation in the broadest sense. Take a long, hard look at

how the company stands in terms of customer service, costs, timeliness, scheduling, accuracy, and marketing plans.

GOAL/QPC in Methuen, Massachusetts, is another organization dedicated to helping propagate the principles of total quality management to smaller companies. It offers seminars, conferences, and a wide range of books and other materials. It translates key Japanese documents on quality and distributes them to its member firms.

GOAL/QPC's 1987 book Better Designs in Half the Time: Implementing QFD in America, by executive director Bob King, was the first book on quality function deployment published in the United States. The organization's memory jogger pocket guides can be customized for individual companies, giving explanations and examples of statistical concepts, along with illustrative diagrams of basic types of charts.

---

involved in quality issues. People become accustomed to making decisions every day when they ask themselves how this will affect quality."

For all the lofty sentiments, the one thing that speaks to small companies loudest is the almighty dollar. Although some of the executives admit to skepticism, all agree that quality programs make financial sense for even the smallest company. At Mips, for example, 3% of the company's $40 million in revenue in 1989 was spent on quality, executives roughly estimate. That level is matched by other quality-conscious small companies, such as tape drive maker Wangtek, which estimates it spends 2 to 3% of sales on quality each year.

↝ The High Costs of Not Doing It Right the First Time

Cadence Design Systems, an electronic design automation software company, has learned that quality programs pay, no matter what the cost, says Mike Macfarlane, group director of software engineering. "There is tremendous leverage for us if we can fix problems in research and development rather than waiting," he explains. "It only costs us $2 to $5 to fix a bug in design, but it's up to $500 during integration and upwards of $1,000 per occurrence in the field."

The payback from quality efforts can be dramatic on the production line, says Jim Hashem, president of Diagnostic Instrument, a small electronic-assembly subcontractor with only about 100 employees. Under pressure from big customers, the tiny company launched a quality program. Although it took time to overcome inertia, particularly with first-line supervisors, the effort to involve everyone in continuous improvement finally began to pay off handsomely. Eighteen months after the program started, Hashem reports: "Business was up 58% over the previous quarter, but the extra load was handled with 15 fewer people and 50% less overtime."

NovaSensor's Ron Ellsworth, director of product assurance, admits: "Our previous decisions used to be based on return on investment. Now quality at all levels is being addressed and any necessary investments are made. The cost of rejects is too high. It saves us money to catch them early on."

Perhaps most compelling is the cost of *not* setting up quality programs. Not having a quality program in place would have cost Cypress Semiconductor 30% of its 1988 revenue of $156 million, estimates Steve Kaplan, vice president for quality and reliability assurance. Other estimates range from 25% to as high as 60% of sales. Spending 2 to 3% to avoid losses of 25 to 30% adds up for even the toughest budget-crunching executives. The loss of future opportunity as a result of defective quality can hurt even more.

"When you're a small company, it's easier, and far cheaper, to satisfy the customers you have than to go out and find new ones," says Richard Oshiro, director for quality at Adaptec, Inc., a maker of disk drive controllers and boards.

Focusing on the customer's needs at the earliest design phases is the approach taken at Cirtek Corporation, which assembles circuit boards for automotive and consumer electronics products. "It's amazing how often design engineers will grab a pistol and start firing before they even know where the target is," comments president David Taylor.

As if small companies weren't busy enough managing their own quality programs, some of them have also found the time to kick around their vendors a bit. "We're not willing to bear the expense of things that don't work," declares Pat Groves, director of operations at computer maker Arche Technologies Inc. Arche carries a 2-year warranty on its own products. To achieve this, it has to be demanding of its vendors—pushing for a 2-year warranty from them as well, even if it means paying a little more.

Novellus takes a different attitude. President Graham sees his vendors as partners. He maintains that adversarial vendor relationships will be the death of any company. In fact, Graham is so willing to work with his vendors—many of which are smaller than Novellus—that he sometimes buys parts for them if they are unable to finance the purchase.

Regardless of the program's origins, small companies are pleased with the success of their programs, although they are constantly striving for better performance. Standing in the way, of course, are such factors as cost and employee motivation, but small companies can point to results that put those concerns to rest.

Sometimes the push for quality can become something of an obsession, as at Sputtered Films and Vamistor, small California and Tennessee firms where chief executives are intent on bagging a Baldrige Award. There are many more examples too of companies that have learned the value of total quality, finding that the payoff can be significant whether they feel they can be a prizewinner or not.

∾ Sputtered Films Targets Highest Quality for Lowest Cost

Peter Clarke, the maverick physicist, inventor, and president of tiny Sputtered Films, Inc., sports a white sweatshirt that boldly invites anyone who approaches him from behind to "Ask me what I've done about quality lately." He's serious. With the fire of one converted, he tells inquirers in great detail about his company's quality journey. He proudly talks about Sputtered Films' crack at the Malcolm Baldrige National Quality Award and describes how the pursuit of quality is transforming the $2.5 million, 22-employee company into the undisputed leader in its narrow field.

The fanaticism isn't hard to fathom: Quality drives customer satisfaction, which in turn drives sales. For the 24-year-old Santa Barbara, California–based company—producer of about six multichamber sputtering systems a year, which are used mainly by semiconductor makers—meeting customer needs is all that counts. Competitors include the $1.3 billion Varian Associates, Inc. and $500 million Applied Materials, Inc.

Since becoming a convert to quality, this minuscule maker of sputtering guns and other semiconductor equipment has made phenomenal strides. It has reduced product cycle time by 75%, cut the cost of rework by 75%, cut customer rejects by 75%, and eliminated final inspection.

Where did Sputtered Films get its quality religion? From Motorola Inc., a 1988 Baldrige Award winner. In June 1989, Clarke received a memo from Ken Stork, Motorola's corporate director of materials and purchasing, informing him that "Any company that does not wish to compete for the Baldrige Award. . . will be disqualified as a Motorola supplier by December 31, 1989."

Clarke's response? "I threw it away." In November, he received a terse letter from Motorola threatening termination of the relationship if a commitment to the Baldrige Award was not forthcoming. This was the clincher. Clarke had an immediate change of heart, and Sputtered Films responded to the challenge with enthusiasm. In fact, Clarke committed the company to compete for the 1990 Baldrige Award in the small company category. Sputtered Films was knocked out in the second-round application evaluations, but Clarke intends to try again. "Applying for Baldrige is the most important move we've made at Sputtered Films in the last 24 years," he insists.

∾ Taguchi Methods Cut Rework Dramatically

The quality crusade at Sputtered Films didn't stop with the Baldrige application. Clarke made all his employees read *Quality by Design: Taguchi Methods and U.S. Industry*, by Lance Ealey. Taguchi methods teach that designing in quality and reliability by definition lowers costs, minimizing rework, returns, and maintenance. In 1989, Clarke claims to have spent less than $10,000 on rework, down from about $40,000 in 1988.

A Sputtered Films maxim is that using economical components and materials while constantly narrowing engineering tolerances ensures the highest product quality at competitive prices. For example, Clarke cut costs by a factor of four by using 6061 aluminum for processing chambers instead of stainless steel. To make the substitution work effectively, the aluminum's surface is treated to match the tolerances of stainless steel.

To help drive home the quality message, Clarke decreed that both employee bonuses and raises would be tied to an employee quality performance index. The index is based on a quarterly evaluation of all employees, even the president, by their 21 peers. There is a personal reward: A reserved parking space, one of only three on the premises, is given to the top quality performer.

The quality commitment shows. In a recent report, VLSI Research, Inc., a San Jose–based semiconductor equipment research house, rated Sputtered Films the top overall equipment maker in the industry in terms of product performance, software support, quality of results, uptime, and customer commitment. Varian, a much larger company that also has a strong commitment to quality, was ranked second. According to Dan Hutcheson, president of VLSI Research, "The difference in Sputtered Films is focus."

⮌ Vamistor: Tiny Resistor Firm Aims for Baldrige Honors

Spend a few hours with John M. Boatman, president and CEO of tiny Vamistor Corporation in Sevierville, Tennessee, and you can't help but get the feeling he's a man with a mission. Despite the overwhelming odds, the amiable Boatman has set a goal for the $1.8 million producer of high-performance resistors and potentiometers to win a Malcolm Baldrige National Quality Award in the small company category.

Achieving that goal, of course, is a dream shared by hundreds of other small U.S. companies. But Boatman feels Vamistor may have a running start on its competition and hopes to be a winner in the future. Vamistor was one of 66 contenders for the Baldrige Award in 1988. Only one award was made in the small manufacturing category that year—to Globe Metallurgical of Beverly, Ohio. No small company won the award in 1989, only Wallace Co., a Texas pipe and equipment distributor, made the grade in 1990, and only Marlow Industries, a tiny Texas maker of thermoelectric devices won in 1991.

Even though Vamistor was unsuccessful in its 1988 attempt, Boatman claims that the exercise of competing for the prize was a priceless lesson in expanding the company's quality consciousness. "It gave us a road map for achieving quality," he says. "We're preparing for our next entry by attacking all the weak areas from '88. They tell you where you fall short, but not how to fix it," Boatman explains.

The feedback from Baldrige examiners to each company that enters the competition can be invaluable in pinning down where efforts must be redoubled, but it takes a willingness to accept criticism and the imagination and drive essential to fixing problems, filling in any gaps in the quality process. In stark contrast to a winner like Xerox, a 1989 winner with a staff of nearly a score of employees that worked specifically on boosting quality ratings, Boatman notes that in 1988 he spent only 10 days on the application and did most of the work himself. For its 1991 target, Boatman planned to set up a small team to support the effort and to get a much earlier start.

Baldrige judges point out both strong and weak points in their feedback. On the subject of Vamistor's product quality, for example, the panel noted that "product reliability is considered superior," and, "the rate of defects per unit is improving." Among downside comments, the judges cited Vamistor's "lack of records on critical customer quality problems." A review of customer satisfaction data, however, indicated that "customers' views . . . are very positive." The feedback report added that "substantiating data relating to customer satisfaction is minimal."

Surveying customers, tracking down any quality problems that do crop up, and keeping good records are all problems Vamistor could set up systems to address. But Boatman admits he does face one serious drawback. While most companies are free to revise product designs and manufacturing processes, and switch to different raw materials and suppliers, his company has no such flexibility. Nearly 90% of Vamistor's business is pegged to military uses. This locks Vamistor into certain processes and procedures—such as 100% testing and inspection of both incoming materials and finished products—that are by their nature at odds with total quality management (TQM). Under TQM, statistical process control (SPC) can help bring production methods under control by cutting down on variation, and sampling can provide evidence of any slippage.

Even with such restrictions, however, Boatman and his small quality team have been plugging away, digging out new ways to reduce the need for inspection and testing without endangering product reliability. "We're required by the Defense Dapartment to provide 100% traceability on every one of our products," explains Boatman. "But at some point the Defense Electronics Supply Command (DESC) will inspect and approve our suppliers' sites. In that case, we may be able to reduce our incoming quality-control checks."

Still, a tour of the plant reveals the tough job Vamistor faces in trying to implement TQM policies. Although assembly of the resistors itself is largely automated, DESC-mandated testing of every finished component is done almost entirely manually and visually. Those processes have resulted in an overall in-house rejection rate of about 16%, according to Boatman, compared with 25% or so in 1989. "We hope to achieve a defect rate of zero by 1991," he says, "but I'm not sure that's realistic in this kind of production environment." To reach that goal, the company designed new test procedures to meet more stringent specs expected from the government calling for new process-control techniques, more efficient equipment, and improved worker training.

∾ Judges Frown on 100% Inspection

Boatman feels that the ultimate measure of Vamistor's quality is the performance of its products in the field. By this criterion, in fact, the company did receive high marks from Baldrige examiners. Still, the panel was not impressed with Vamistor's reliance on 100% inspection rather than on the implementation of cost-effective process control. Instead of striving to prevent production errors—a fundamental precept of quality assurance—the firm depends on a relatively large number of workers to spot problems.

The lack of an adequate SPC system cost the company points, and even more were lost because of the seemingly marginal involvement of the entire staff in the quality process. "At the time we were just getting started in SPC and our data base was quite small," admits Boatman. "We also were in the very early stages of introducing the entire company to the concepts of total quality assurance. Today, we're much more structured and formalized."

Vamistor is steadily tackling each of these problems. Even where regulations restrict its flexibility the company is taking steps to improve where it can. For example, new methods are being designed for soliciting and reviewing suggestions from the production staff for reducing product inspection without degrading product performance.

Companies committed to a serious run for the Baldrige Award must be in it for the long haul, as Vamistor's efforts illustrate. "I think the biggest obstacle for us is the need for continuous worker training," says Boatman. "We have to constantly impress on our employees how important every part is."

Supplier relations is another sticky area. The company has built favorable long-term relationships, according to Larry Simonds, quality assurance manager. "But I'd like to see more competition between our suppliers. The number of domestic sources for ceramics and leads is shrinking. That gives those that remain a lock on the market, so we have to take what we can get at their price." Although Simonds has not seen evidence of declining quality of purchased materials, he considers price—along with delivery and product integrity—an important element of total quality.

What galls Boatman the most, it seems, is the conflicting messages coming from the Pentagon. "They're pushing the idea of total quality," he says, "while at the same time stating their intention to use commercial systems whenever possible. Our whole purpose is to supply military-quality products. So do I now invest in the equipment I need to increase quality even further?"

Faced with a limited product line-up and impending cuts in defense spending over the next few years, Vamistor is looking to increase its business in commercial markets, such as industrial and highway lighting fixtures and high-voltage power supplies. Although commercial sales now make up only about 12% of Vamistor's business, the company believes there is potential demand in the marketplace for products designed to meet tough military specifications.

Meanwhile, winning a Baldrige Award couldn't do any harm!

～ Small Company Quality: How the Little Guys Do It

Following is a sampling of quality programs at some other small companies across the country.

～ Diagnostic Instrument Corporation (Ayer, Massachusetts)

Diagnostics Instruments Corporation, a tiny manufacturing electronic-assembly subcontractor, with 100 employees and 1988 sales of $5.1 million, launched a quality program as a result of pressure from its big customers, particularly Digital Equipment Corporation President Jim Hashem now says he believes quality control is not just a good idea, "it's necessary for survival." But implementing a program can't be delegated, he believes. "Top management has to get their hands dirty."

Hashem and eight of his supervisors started a continuous improvement process based on W. Edward Deming's principles after attending seminars offered by GOAL/QPC, a consulting organization in Methuen, Massachusetts, held two afternoons a week for eight weeks in the fall of 1987. Soon thereafter he hired Leo Nangle, who had been a quality engineer for EG&G, Inc. in Salem, Massachusetts, as quality control manager.

Hashem says it took plenty of reassurance from top management to get the quality process rolling. But the results have been worth it. In 1987 only 70–75% of boards were defect-free after first-pass production. Even though the company has toughened its definition of defects, by 1989 95% of its boards were good on the first pass. "Doing things right the first time really pays off," Hashem says. Since the company operates on fixed prices, "we pay for our own mistakes."

～ Mips Computer Systems Inc. (Sunnyvale, California)

With the idea of exceeding its customers' expectations of quality, Mips has developed specific quality groups for each of its major product lines, according to manufacturing vice president Tom Rohrs. Now Mips has teams of people to analyze problems. "The result is that the quality of our outgoing product has increased dramatically and our cost structure is lower," Rohrs says. "High-quality process and high-quality design yield a lower cost product."

Being small, Mips avoids the communications problems that plague larger companies. "Enthusiasm and energy hit high levels here," he explains. "It's easier to fire them up, and management is closer to the action."

The point Mips employees rally around most is the cost of not producing quality, which Rohrs estimates is around 30 to 35%—a big penalty, he says, for not doing it right the first time. "It's not just an issue of throwing away a bad part. You need to work hard to remove waste in all your processes," he stresses.

～ Cirtek Corporation (Flint, Michigan)

Manufacturing exactly what customers want and need has occasionally been a problem for Cirtek, a high-volume circuit board assembly operation, especially with such demanding customers as automotive and consumer electronics companies. The company has about 250 employees and did $25 million in business last year.

"We had to learn what Japanese companies are so good at: listening to the voice of the customer," president David Taylor explains. Cirtek worked with Ford Motor Company to learn the principles of quality function deployment. Now the company is serving as a beta-test site for new software that generates charts to help designers focus their efforts on customer needs and prioritize their work, which Ford will soon be marketing to other companies.

"If we were as good 5 years ago as we are now, we'd have blown all our competitors out of the water," Taylor says. "But now even defect rates of a few hundred parts per million are unacceptable." To boost quality even higher, the company hired an outside consultant, a Michigan State professor, to act as a facilitator for systems improvement groups in the factory.

"We are also switching from a push to a pull system in our plant," Taylor says. Workers use walkie-talkies at each work station to call for new work when bins are nearly empty. This will greatly reduce in-process inventory, Taylor believes, and will also boost quality by speeding the flow of work through the plant. "There's much more time for trouble to happen if a product lies around the plant for a day or two rather than going through production in 2 hours," he says.

～ NovaSensor (Fremont, California)

Silicon pressure sensor company NovaSensor is trying to implement SPC and quality function deployment (QFD) in an effort to increase its quality. The number one reason: it is currently going after customers in the automotive industry, and a more quality-sensitive bunch would be hard to find.

It's not easy, product assurance director Ron Ellsworth says. "We need to grow to a certain size to fill all the holes we want to fill." Two years into the program, NovaSensor was spending 5% or more of its revenue on quality.

Too often in the past in U.S. industry, manufacturers were satisfied when they were just able to meet customers' specifications. One way that Nova cuts defects is to improve processes continually, until the average results are five times better than are called for in the specs. Ellsworth feels that managers in Western countries are still too focused on detection of problems rather than prevention at the source.

He uses an analogy from baseball to convince workers of their importance in finding the sources of trouble and fixing them: An average pro hitter has a batting average in the range of 250 to 280, and an outstanding one hits 300 to 370. Yet a team average of 260 to 270 wins the pennant 60% of the time, while a 280 to 290 team average wins 95% of the time. That's why, Ellsworth points out, if everyone improves 5%, the overall result for the team might be 50% better.

In general, the company's yields have increased from 92% to 98% at the final inspection stage. But that's still not good enough. "We need 3 to 5 years to complete the implementation of this program," Ellsworth says.

～ Vicor Corporation (Andover, Massachusetts)

Power supply maker Vicor "recognized early on that our growth would be sustained by manufacturing quality," says Jim Muckenhoupt, director of marketing. So in 1988 the company put in statistical process controls to monitor the assembly process. The result: in-process yields went from 40% in 1988 to the mid-80s in 1989. Final product yields are in the high 90s, quality assurance director Charles Ackerman reports.

Vicor executives view quality as a process of continuous improvement, Muckenhoupt says. It also has to be a real program, not just one on paper, he stresses. "It's not a marketing thing. People, customers, see through that right away."

The company's emphasis on quality even extends to the process of hiring. Hiring is a team decision, Muckenhoupt says. "There is a general feeling of team participation. We want committee-type approval. It's expensive, but relative to the cost of hiring the wrong person, it's minimal."

～ Cypress Semiconductor Corporation (San Jose, California)

After some aborted attempts at boosting quality, Cypress president and chief executive T. J. Rodgers decided enough was enough and brought in a consultant to help the start-up put a true quality program in place. As a result, Cypress established quality improvement teams which target specific problem areas within the company. Mid-level managers lead these teams, and executives all the way up to Rodgers himself participate, says Steve Kaplan, vice president for quality and reliability assurance.

Kaplan, after only a few months in this newly created post, says the team efforts are starting to pay off.

In new product development scheduling, for example, Cypress used to miss 50% of its targets, Kaplan says. Since implementing quality teams, only 5 to 7% of the deadlines are being missed. With steady improvements across all its operations, Kaplan says Cypress is committed to going after the Malcolm Baldrige National Quality Award but expects it will take a few years to get there.

∼ Wangtek Inc. (Simi Valley, California)

Wangtek pursues quality primarily by "making sure we catch the problem before it goes out into the field," says Rick D. Roberts, director for quality assurance. A maker of tape drives, Wangtek spends time and energy on establishing a system of problem tracking and solving. The company also monitors field data extensively, again with an eye to quality.

Wangtek tries to leverage its strengths in quality and play down its potential weakness, which is its size. "Small companies are forced to do some things smarter," Roberts says. Wangtek hopes to catch problems earlier by employing a smarter work force. To get there, the company brings in experts for 1- and 2-week training sessions, both in the classroom and on the job.

These are just a few of thousands of small companies that are developing their own unique approaches to total quality. As the message propagates from the corporate giants through tiers of suppliers and subcontractors, any small company interested in staying in business would do well to heed the call to action.

Chapter 13

Computer Companies Vie for Quality Leadership

Many computer companies, facing growing marketplace pressures just like market-leader IBM, have developed their own formulas to become world-class competitors. Compaq Computer in Houston is not only the most successful builder of IBM-compatible desktop and portable computers, it is also challenging IBM in the total quality arena. Digital Equipment Corporation in Maynard, Massachusetts, the long-time leader in minicomputer systems, has developed its own approaches to total quality management (TQM). Upstart Next, maker of a futuristic desktop computer system, illustrates how even newcomers to the industry can challenge the leaders in manufacturing quality as well as with innovative technology.

Because both the technology and the marketplace for computers are so fast paced, companies like these feel they must get out in front in the quest for total quality. That's why their stories should prove helpful to any organization, no matter how advanced in TQM methods.

⁓ *Compaq: Reaching the Top Begins at the Bottom*

In competing with IBM for the high ground of the personal computer market, Compaq Computer recognizes the importance of quality. Unlike microcomputer clone manufacturers that compete largely on price, Compaq competes in the personal computer market with products that are equal to or better than IBM's in terms of performance, functionality, and reliability. Compaq, in short, is selling quality.

The formula is working. In 1990 the company's sales topped $3 billion. Compaq consistently ranks high on customer satisfaction surveys. In a 1988 survey conducted by Techtel Corporation, an Emeryville, California, market research firm, Compaq achieved a customer satisfaction rating of 95%. "Clearly we would not have been able to achieve our market share and customer loyalty if we had not had quality products," says Michael S. Swavely, former president of the company's North American operations.

To maintain that edge, Compaq has adopted many of the same quality control programs that are used by other electronics vendors, including statistical process control, vendor certification, and total quality control. The difference at Compaq is how those programs are implemented. While many companies use a version of the trickle-down theory and dictate quality from the top, Compaq takes the low road, implementing its quality programs instead from the bottom up. "The difference is the way that we apply quality control and the way we involve people in it," says Bob Vieau, vice president for corporate manufacturing.

The quest for quality has "permeated every aspect of the organization," adds Murray Francois, senior vice president for corporate quality and materials. "It's woven into the very fabric of the company." As of 1990, Compaq had not applied for the Baldrige National Quality Award but was considering a future entry.

At Compaq, most departments operate their own quality programs, according to Francois. He oversees the individual programs—in the case of manufacturing, for example, by conducting post-sales quality checks of products—and develops training programs for each department. He is also in charge of ensuring quality of materials from the company's suppliers.

Total quality management, or total quality commitment, as Compaq likes to call it, makes up the core of the Houston company's quality programs. First widely applied in the 1950s in postwar Japan, where reducing waste was a priority, total quality management has been adopted by Compaq and other U.S. companies by molding its methods to their own corporate cultures.

Total quality management provides "an operating philosophy and a language which everybody can understand," according to Vieau. He says that Compaq has been training all of its employees in the procedures, including those in materials, facilities management, and human resources. Vieau says that the first step is to simplify the process, whether that process is manufacturing, hiring new employees, or some other company operation. Phase two, which goes hand in hand with the first, is to use the simplified process to eliminate waste. The third step involves instilling "a philosophy of continued improvement," in employees, according to Vieau.

The goal of the final step of total quality management is to understand customer needs and meet their expectations. "Customers" does not necessarily mean Compaq's end-user customers, but the person or group for whom an employee performs a task. The personnel department's customers, for example, are those internal operations for which personnel hires new workers.

～ Measuring Progress Is Critical

Vieau says that Compaq conducts its quality programs by establishing goals, communicating those goals among employees, and measuring progress toward meeting those goals. The measurement process requires progress reports and feedback, both up and down the chain of command.

Vieau cites an example involving human resources. The department had been filling other departments' hiring needs within 90 days. But in a meeting with their "customers" (other department heads), the human resources department learned that the hiring process needed to be shortened to 60 days. "The methodology is staying in step with needs and feedback on how you are performing," says Vieau.

Perhaps nowhere are Compaq's efforts to achieve quality from the bottom up more visible than in its manufacturing operations. Although Compaq has always involved its managers in quality control, the company extended the program to workers—using a team approach—in July 1987.

In 1985, clone personal computer manufacturers were invading the market and putting pressure on Compaq's profit margins. Vieau says that this forced Compaq to look closely at its cost structure and decide how it would build its products in the future. Like other computer manufacturers, Compaq considered automation as a way of trimming its expenses. It looked at the auto industry as a possible role model for automation, according to Vieau, what they found was that automation had not solved the auto industry's quality control problems. "I think what they found out is that they automated a bad process," says Vieau.

He believes there is a difference between automation, which he sees as a way of improving the efficiency and quality control of a manufacturing process, and mechanization, which he believes is simply a method of cutting labor costs. Francois says Compaq is striving for quality and flexibility, not to just cut costs. "We haven't overmechanized," he says. About 2,600 of Compaq's 8,500 employees worldwide are directly involved in assembly and test operations, according to Vieau. He says that labor and overhead as a percentage of total costs is less than 10%, compared with the industry average of 12 to 15%.

Compaq has become highly automated in its printed circuit board manufacturing, an operation that involves a large number of repetitive tasks. Vieau believes that quality can suffer when such tedious jobs are performed by people, and so automating those operations provided a gain in quality. In 1988 Compaq also began using statistical process control in its board operations.

∼ Worker Flexibility Aids Productivity

Compaq's view toward its assembly and test operations is that while many tasks could be performed by robots, people are more flexible. Shop floor employees are urged to make changes and resolve problems "at the individual operator level," Francois says.

Most manufacturers can resolve big problems, the kind of snafus that shut down entire production lines. But Vieau says that most companies also suffer numerous chronic problems that, while not enough to bring manufacturing to a halt, add to the costs of production.

Vieau says that those kinds of problems can only be solved by people at the assembly line level. To achieve this, Compaq gives its assembly line workers an extraordinary amount of decision-making leeway and opportunity for input. Total quality teams, for example, can jointly decide when to start and end their shift.

A tour of the manufacturing area turns up other examples. Assembly line workers are never far from pull-cords that allow them to halt a production line if they feel that something is amiss. Around the floor, easels are set up on which workers write complaints, comments, and ideas. "Unseated shock mount," states the finding of one inspector. "Slow response from Engineering!" gripes another.

Workers are encouraged to innovate in order to solve a problem. One of many employees who inspect systems as they pass by on the line came up with the idea of positioning a mirror so that he could easily view both sides of the product at once. The idea caught on, and now a number of inspectors use the same technique.

Assembly line teams, averaging 10 to 20 people in size, meet at the end of each shift and discuss what went on during the day and what problems were encountered. The workers identify what problems they can fix themselves and what problems require outside help. Each month the group leader prepares a report on the results of the team meetings.

Before the total quality team approach, according to Vieau, the attitude of employees was: "Somebody needs to go fix that problem." He says that today workers are solving problems that they did not even know they could be responsible for, let alone resolve on their own. Vieau says the concept of self-managed work teams may be expanded even further, allowing the teams to make changes in the manufacturing process and perhaps even to determine their own hiring needs.

Manufacturing is traditionally viewed as a problem in such fast growing areas as the personal computer industry, according to Swavely. But he says that Compaq's flexible and responsive manufacturing operations provide the company with a strong marketing tool by allowing the vendor to respond quickly to customer demands.

◡ DEC: Digital Strives for a Consistent Vision of Quality

Digital Equipment Corporation is no stranger to quality improvement, having introduced aspects of total quality management within the company as far back as 1983. But the computer maker, based in Maynard, Massachusetts, formally launched a corporate TQM program early in 1990 to tie together the efforts scattered throughout the company.

"With the TQM model we launched in 1990, we are trying to carry a consistent vision or language throughout the whole company," says Ken Potashner, group manager for corporate quality and technology. The program includes a 6-year journey to six-sigma, a statistical measure equivalent to 3.4 defects per million operations. With such a framework, decentralized Digital expects to find it easier to apply the quality-improvement experience and benefits gained in one operation throughout the company.

Digital is also motivated by its need to improve its competitive position within the computer industry. "It is absolutely in response to a need to intensify our competitiveness," Potashner agrees. While he says that customer satisfaction has always been the goal at Digital, the new initiatives will help the company meet that objective at competitive costs.

The new program is made up of four initiatives:

• Reach six-sigma quality by 1996. Using the program developed by Malcolm Baldrige National Quality Award winner Motorola, Digital seeks to eliminate waste and defects throughout its operations at a rate of 60% a year.
• Listen to the voice of the customer, thus ensuring that every task within the company is performed with customer requirements in mind. "It goes well beyond customer surveys," says Potashner. "It is trying to live in the customers' shoes." The approach uses methodologies like quality function deployment, a system for determining what is important to the customer and what the alternatives are to meet those needs.

∼ DIGITAL'S INITIATIVES FOR QUALITY IMPROVEMENT

- Voice of the customer: Ensures that every task within the company is performed with a customer requirement in mind
- Benchmarking: Determines what company is the best at a particular task for use as a role model
- Six-sigma: Eliminates waste and defects in every element of work within the company
- Improve cycle time: Determines actual and optimum times needed to complete a process and establishes a plan to reduce the cycle time

- Use benchmarking techniques to determine the leader in a particular operation, then create a program or educational curriculum to match its performance. Role models may come from any industry; outdoor-clothing vendor L.L. Bean is an example of a successful distributor. "The essence of benchmarking is that we go beyond our industry," Potashner says. Digital is working with Xerox Corporation (also a Baldrige Award winner), Motorola, and Boeing Co. to establish a benchmarking program;
- Improve cycle time. Digital determines the theoretical minimum or optimal cycle time, the actual time now needed to complete a process, and what has to be done to change from current to optimal cycle time. The program eliminates unnecessary steps in manufacturing, procurement, and other processes.

Previously, quality programs largely focused on manufacturing, especially on cycle-time reduction and continuous improvement. For example, Digital cut cycle times in half for its low-end system using product simulation, design for manufacturing, and concurrent engineering techniques. The RA-90 disk drive, introduced in 1988, has a mean-time-between-failure rate twice that of its predecessor, RA-82.

Several individual units within Digital have become the spearheads for TQM efforts, including Unix systems and services group, telecommunications, and PC systems and peripherals. The last group, under the direction of vice president Grant Saviers, has played a key role in bringing six sigma to Digital and was the first of the company's operations to inaugurate a trickle-down training process the company calls "cascade training."

Digital is still evaluating where it stands on the six-sigma scale. Potashner says that different operations within the company have achieved different levels of quality improvement, but he declines to reveal those statistics, arguing that the methodologies are more important than the statistics.

The company takes both a bottom-up and a top-down approach to quality improvement—bottom-up in that programs are implemented within all operations, top-down in that executives ensure that quality improvement efforts are consistent with company goals. "The vision and pressure for consistency comes right from the top," says Potashner.

Frank McCabe, quality and technology vice president, is in charge of Digital's quality improvement programs. Potashner, previously a technology manager at the company's Ireland operations, is responsible for the six-sigma program.

Through education and such programs as small group improvement activity (Digital's version of quality circles or problem-solving teams), employees are given the tools that management hopes will allow them to make quality gains on their own. "The whole initiative sits on a platform of an involved and an empowered employee base," says Potashner.

The problem facing corporate education manager Rebecca Raibley was how to educate DEC's 120,000 employees in quality improvement and yet provide a consistent curriculum throughout the company. The answer was a train-the-trainer program and the trickle-down effort Digital calls cascade training. "From a strategic view, DEC is so decentralized that it doesn't make sense to have centralized training," says Raibley. "It's best to develop a curriculum that is implemented locally." Besides the huge expense of having a centralized educational system, Raibley notes that local sites need quality improvement instructors on a day-to-day basis—a difficult task to accomplish given the size of the Digital empire. "You really need to have people there who have the skills, tools, and knowledge."

Digital began its cascade training in 1990 with 1-day workshops, introducing all employees to the basics of total quality management and the company's four-point quality improvement initiative. The goal is to put all 120,000 employees through the workshop.

Employees who want to become trainers must take a basic problem-solving techniques course, covering quality improvement tools and methodologies, which is taught locally a half day a week for 10 weeks. That is followed by a more intensive 3-day course in problem solving, a week-long course in teaching skills, and training in such specific methodologies as process management and quality function deployment. Prospective trainers must practice in front of a video camera, pass a master exam, and co-teach the 10-week course with a qualified trainer before becoming certified instructors.

To date, 163 Digital employees, ranging from senior managers to manufacturing floor operators, have been certified as instructors, although the emphasis of late has been on middle managers. Under those teachers, some 50,000 Digital employees have undergone some form of quality improvement training since the program began in 1983.

∽ Emphasizing the Need to Improve

One task now facing Digital is the need to create an atmosphere that encourages employees to be honest about their need to improve. "An individual must be able to stand up and say 'I am 50% off the benchmark, but here is my improvement strategy,'" says Potashner. "The trust that needs to be established is one of the absolute foundations of this total quality system."

Suppliers are also required to be committed to continuous improvement, but Digital allows them a great deal of flexibility in the methods they use to meet product quality criteria.

Digital also hopes to boost its quality efforts through its participation in the Center for Quality Management, a consortium of 23 companies that share knowledge and experience in implementing quality improvement programs. Associated with the Leaders for Manufacturing program at the Massachusetts Institute of Technology, the consortium was founded in 1990 by Digital, Analog Devices Inc., Teradyne, Inc., and other companies.

After becoming a finalist for the Baldrige Award in 1988, Digital decided to hold off reapplying while it digests what it learned in that first contest and gets its quality house in order. "We felt we had plenty to work with," Potashner says. Don't expect Digital to hold off reapplying for very long. Plans are to take another run at the coveted trophy as soon as its ducks are in a row.

✕ Next: Manufacturing Is Its Quality Cornerstone

While some sectors of American industry have neglected the manufacturing side of their business, the computer industry has been one area where the looming threat of Japanese competition has forced management to focus on production skills. Most computer start-ups, however, have used innovative design as their market entree and only later tried to bring their manufacturing up to world-class standards. Next Computer Corporation is an outstanding exception to the rule.

With $133 million in total financing, Steve Jobs, one-time cofounder of Apple Computer and now Next's maverick CEO, spared no expense in creating the factory of the future even as the firm's futuristic desktop machine was coming off the drawing boards. In a brand-new facility in Fremont, California, Next put together a production line so advanced it is the envy of even the Japanese. Jobs's rationale is simple: A highly automated quick turnaround manufacturing process produces the highest quality products at the lowest cost.

In operation since October 1987, the state-of-the-art factory has wowed those privileged enough to get through the heavily guarded security gate. Says Martin Piszczalski, senior industry analyst with the Yankee Group, a Boston-based market researcher, "Steve Jobs has laid the groundwork for one of the most efficient plants in the world."

Even though sales of Next's sleek, black workstation have been lackluster, the capabilities of the manufacturing plant are first rate. The company claims its $20 million facility can produce $100 million worth of workstations a year with only four assembly-line workers. "Other companies would have to have 100 people to do that," boasts Randy Heffner, Next's vice president of manufacturing. In addition to the four employees on the line, two employees work in automation, two in testing and four in final assembly, for a total of 12, says Kim Spitznagel, manager of automation.

The company did not implement the extensive automation of the manufacturing facility to reduce personnel costs, says Heffner. Next made most of the decisions on the manufacturing organization, including the intensive automation, to establish and maintain consistently high quality.

Next's emphasis on quality prompted the company to apply for the Malcolm Baldrige National Quality Award in 1990. Next was not given a site visit by the Baldrige examiners primarily because of the lack of historical data, according to Next officials. The company plans to try again, however.

In Next's case, high quality is tightly coupled to productivity. Next's plant could approach $10 million in sales per production worker, says Piszczalski, a number that is unparalleled by any U.S. or Japanese electronics company. Average productivity rates in the electronics industry run anywhere from $80,000 to $400,000 per employee, the Yankee Group estimates.

Next's advantage in manufacturing has been its ability to start from scratch with virtually unlimited funds. Unlike the situation at most struggling start-ups, Jobs brought prestige and leverage to Next. He invested $7 million of his own money in 1985, then brought in $26 million from other sources in 1987, including superstar industrial investor H. Ross Perot. In 1989, Canon, Inc. of Japan chipped in an even $100 million.

To Jobs's credit, he recognized early in the start-up phase that manufacturing was as fundamental to the company as product design and marketing prowess. To this end, Next built the 40,000-square-foot facility.

"Most start-ups have a money-man, somebody to design a product and a marketing person," says Heffner. The marketing person gets orders, then the company goes looking for a factory to make the product.

"By contrast, at Next, time and energy were put into designing a manufacturing system from the outset that would allow great flexibility," says Heffner. That meant hiring top-notch manufacturing people. Heffner came to Next from Hewlett-Packard Co. and Spitznagel from Motorola. Next executives point out with pride that 70% of the manufacturing staff—38 people in all—have Ph.Ds. That is a far cry from the average electronics manufacturing company, which some analysts say has fewer than 70% of its manufacturing staff holding high school degrees. Equally important to Next is the fact that the manufacturing staff is as well compensated as the research and development staffs.

Next holds an unconventional view of the factory, in part because of the educational level of its staff. "We're more and more looking at manufacturing as an interesting software challenge," Jobs said in a recent speech. The commitment to that philosophy is confirmed by a Yankee Group report. Most of the workers in the plant have software expertise, according to the report, and all the machines are treated as computer peripherals. To the staff, operating the plant is like running a big computer. "We can compute in 20 minutes what used to take 3 weeks to calculate. And we can start building boards 20 minutes later. Software development has saved us a tremendous amount of time and has increased our quality tremendously," Jobs said.

Much of that productivity- and quality-enhancing software resides in the factory-floor workstations. Operators can quickly perform statistical process control and track both process-performance and defect rates.

Almost all plant software applications are internally developed, with Unix being the operating system for the entire plant. Finance, sales, order entry, and other management information systems functions have been centralized.

The software skill set on the factory floor has allowed Next to finely tune the manufacturing process. For instance, the plant staff wrote the machine-control and process-control software. To ensure design for manufacturability, staffers set up a software capability to transfer data quickly from design engineering to manufacturing. They also wrote software to make CAD data easy for the factory-floor robots to use. Next uses Hewlett-Packard workstations, including those from Apollo, and graphical-

∽ Four Keys to Next's Productivity

1. a highly educated work force
2. the use of nine Unix-based workstations, the workhorse computers of the plant
3. graphical user interfaces for all plant workers
4. computer-aided software-engineering tools for rapid modification of the computer-integrated-manufacturing system

user-interface software, particularly Next's own Next Step and packages from Mentor Graphics Corporation

The networked software helps streamline the design process to the point where it takes only about 30 days to tool up a totally new circuit board, according to Next executives. The regular manufacturing line even builds prototypes. The production line is so thoroughly automated that it takes about half an hour to switch over at the end of a day from regular production to building a prototype, Heffner says.

Although the Fremont plant had been dedicated to one product—a computer workstation with three memory options and four choices of disk drives—flexibility will also allow additional products to be manufactured on the same line. Next makes the computer with only one central processing unit and installs about 85% of the components using surface-mount technology. The actual manufacturing area takes up 26,000 square feet, but the factory was designed so that additional capacity could be brought on line quickly, says Spitznagel.

∽ Cutting Components for Reliability and Manufacturability

Another factor in Next's quality push is design of a system to minimize the number of components used. "We changed 21 different parts to improve quality and reliability in the design of the Next computer," Heffner recalls. Some parts were completely eliminated. "After all, the most reliable part is one you don't have to use."

Although Next has not accumulated many benchmark statistics, Heffner believes that various production procedures show significantly fewer defects than the industry norm. For example, the solder-joint defect rate at Next averages 4 to 6 defects per million, according to Heffner, who estimates that 200 to 350 defects per million is the industry average. The Next computer has 465 components, requiring 3,300 interconnects, Heffner says, much less than the industry average.

The usual approach to component qualification is to allow the research-and-development staff to source the components with the suppliers. "R&D usually enjoys the honeymoon with the supplier," says Heffner—that is, they work with the supplier for the first 3 to 6 months until the part is satisfactorily designed. "The person who then has to live with the choice is me, the manufacturing guy. I'm the one who has to stay in the marriage for years."

At Next, suppliers must work from the start with manufacturing. To keep tight control on quality, Next keeps its vendor list short. The company deals with only 20 key suppliers and 45 suppliers in all.

Because about 90% of performance characteristics in a part are of more concern to manufacturing than to research and development, such issues as packaging or how a component can be lined up to work with automated equipment should be the purview of manufacturing, claims Heffner. Involving manufacturing in design issues is a radical departure from the usual methods of U.S. manufacturers. But then Next is blazing new trails that could well set the pace for others to follow.

Chapter 14 ⟿
Military Contractors Push for Quality

With huge procurement budgets, the military services exert a powerful influence on American business and industry. Although the Pentagon has professed support for the principles of total quality management, the practical reality has been more a series of fits and starts than a consistent, effective program.

One problem has been the turnover of top policymakers and procurement executives. Top echelon changes usually entail a shift in emphasis that sends new directives bubbling through the massive military procurement hierarchy, often with each service giving its own twist to the overall policy direction. Industrial contractors, not surprisingly, tend to react cautiously, recognizing that just like many programs of the past, the new approach may also rest on shifting sands.

Perhaps even more of a hindrance is the Pentagon's traditional approach to quality assurance. The procurement process requires stringent adherence to military specifications, followed by testing to ensure compliance. This is the antithesis of a true total quality approach. An underlying theme of TQM is that suppliers must totally understand customers' needs and then adjust their own systems and processes to meet these requirements as effectively as they can instead of blindly following specifications. To continually work toward this goal, a manufacturer must track down causes of variation and defects and make adjustments to eliminate them. But the mil-spec approach often precludes changes, so even when contractors develop TQM methods in their commercial business they find that they can not apply the same benefits to military work.

This is ironic, since some experts trace the rise of current methods of total quality and statistical methods to the work of the U.S. Department of War during World War II. Two fathers of modern statistics, Sir Ronald Fisher and Walter Shewhart, laid the groundwork earlier in this century, and their work was later refined by W. Edwards Deming. The use of their statistical and quality control methods was considered so critical to the war effort that they were classified as military secrets—known as "Z-1" in the United States and "Standards 600" in Great Britain—until after the war. Even more ironic, it was the U.S. occupation forces in Japan that ordered the defeated nation to apply statistical methods to rebuild its telecommunications industry. The rest is history.

After a number of attempts to find ways to move defense contracting more toward the total quality model, the Pentagon recently moved toward a total quality management program. A key element is development of a qualified manufacturers list (QML) through on-site evaluations of contractors' plants. Another program would send failed parts back to contractors along with an analysis of the breakdowns, helping vendors to trace root causes. Meanwhile, each of the military services is developing its own variations on the total quality theme. Military budget cutbacks, along with the ponderousness of the contracting bureaucracy, have slowed the transformation, and observers believe it will take years to see if the seeds now being planted take root.

While the Pentagon's own total quality program sputtered into existence, many military contractors devised their own efforts. Some of these stemmed from internal troubles or from management looking for improved performance over the long haul rather than just short-term results. Signs of total quality stirrings within the Pentagon undoubtedly spurred some industry programs. Following are some examples of how specific military contractors are implementing and expanding their own quests for total quality.

∾ Texas Instruments: Defense Group Pushes Quality to the Front Lines

Managers at Texas Instruments, Inc.'s (TI's) Defense Systems and Electronics Group (DS&EG) hesitated before applying for the Malcolm Baldrige National Quality Award in 1990. With defense contractors looking to the commercial sector for guidance, could a defense contractor meet the quality standards being set in the commercial market?

"It's probably no secret that the defense industry doesn't have the best quality track record in the world," acknowledges John P. Leslie, total quality planning manager for the Dallas-based defense group. One reason TI's defense group decided to go for the award is because the company's executives believe their quality improvement efforts surmount the defense industry's poor reputation and can provide a role model. "Certainly," Leslie concludes, "others can learn from us."

DS&EG, for example, helped the Department of Defense establish its own quality criteria. The Pentagon designated the unit's Lewisville, Texas, missile-assembly plant an "exemplary facility," and the Air Force Systems Command product-assurance program sent its own managers to TI to attend the company's quality improvement workshops.

"It's interesting that when we first started to look at Baldrige, it looked very familiar to us," says Leslie. That is because in both 1988 and 1989, according to Leslie, each function within the defense group measured itself against the Baldrige criteria even though the division did not enter the competition. Those functions included the weapon systems, electrooptics, and avionics-systems entities; the business-development and marketing organizations; the personnel, control, and contracts departments; and the quality, reliability, and operations organization.

The information gathered from those two trial runs formed the foundation of

DS&EG's Baldrige application by helping to reveal any weaknesses in the group's quality programs. The Baldrige Award is not seen as an end in itself but simply "the next step in our quality journey," says William B. Mitchell, formerly DS&EG president and a TI executive vice president. "It's not a stopping point. The idea of continuous improvement must be understood by everyone from the top to the bottom. It's a continuous learning experience." That's why the group used feedback from the 1990 competition to try again in 1991.

TI's quality programs revolve around what the company calls its three cornerstones: customer focus, employee involvement, and continuous improvement. Rather than viewing its quality initiative as an exercise unto itself, TI incorporates the cornerstones into its business strategies, tactics, objectives, and management styles.

Managers, for example, are exhorted to keep quality in mind when drawing up long-range business plans, says Mike Cooney, vice president and the division's manager of quality and reliability assurance. Another TI goal is what Cooney calls "cross-functional integration"—expanding quality improvement efforts beyond isolated pockets in manufacturing and design. The premise for cross-functionality is straightforward: Quality can't be achieved in manufacturing if the original product design is bad. Also, by improving the testability, producibility, and reliability of a product early in the design stage, TI is speeding its time to market.

⟆ Quality Cornerstones Lay the Foundation

"What we are trying to do is not only teach quality but show people, managers included, how to use it in their day-to-day decision-making," says Jimmy C. Houlditch, formerly senior vice president and manager of quality, reliability, and operations. Customer focus at TI is embodied in its quality policy, copies of which are posted throughout the company's facilities: "For every product or service we offer, we will understand the requirements that meet the customers' needs, and we will conform to those requirements without exception." TI employees adopt the philosophy that everyone has a customer, internal or external, whose expectations must be met.

"We've really got to understand the total requirements of our customers so that we can meet those requirements," says Mitchell. To that end, customers are brought into TI plants to give workers direct feedback on their work. Curt James, a weapons-systems marketing engineer at the division's Lewisville facility, describes how one Navy pilot told employees that the TI-built HARM high-speed antiradar homing missile functioned without a single failure throughout a 7-month aircraft carrier mission. Such meetings, called "all-hands briefings," also serve to boost employee morale, says James.

TI also sends its people into the field to see defense products in use. James tells how TI personnel, by watching Navy maintenance crews handle the division's missiles on the pitching deck of an aircraft carrier, learned firsthand the importance of making weapons systems easy to work with and maintain.

Even workers are constantly reminded who their customers are. At the Lewisville plant, posters with photos of Air Force pilots bear the powerful message: "The next missile you build may be the one that saves my life."

"If you're going to reach the objective of customer satisfaction, you've got to work with your customer well before the design phase," says Cooney. "You've got to understand your customers, understand what their needs are and have the right technology solution to meet those needs."

Texas Instruments involves employees in a number of ways. Perhaps most visible is TI's annual employee survey, which asks workers to comment on issues ranging from cafeteria food to attitudes about career-advancement possibilities. One question, asking whether the employee understands the quality requirements of his or her job, was answered in the affirmative by 93.1% of all respondents in the 1990 survey.

DS&EG also stresses ethics as a fundamental component of its quality programs and conducts ethics in business courses for salaried personnel (totaling 16 hours) and hourly workers (totaling 4 hours). The training covers such subjects as dealing truthfully with customers and the employees' obligation to report errors or misdeeds.

Throughout the 1980s, the defense group instituted quality-training programs for its personnel and adopted many of the quality-improvement methods being devised for design and manufacturing operations. The group, for example, began using statistical process control in 1984 and adopted Genichi Taguchi's design of experiments methodology in 1988.

In 1984 the division formed effectiveness teams, its own version of quality circles. Each is organized according to discipline, such as assembly or test, and provided with the means to make improvements in operations. For example, a plant-floor supervisor in the Lewisville assembly area proudly showed off a computer-driven device, developed by his effectiveness team, that automatically marks bad solder points on printed circuit boards that are destined for use inside HARM missiles.

Cooney says the division is also beginning to form cell teams, which combine personnel from various disciplines. Cell teams look at multiple operations, such as design, assembly, and test, with the goal of better integrating those operations.

∼ TI Sends Managers to Quality Classroom

Texas Instruments' decision to send 600 managers to the Crosby Quality College in 1981 provided the spark for the Defense System and Electronic Group's quality efforts. "That planted the seed and gave us the fundamentals to establish this quality thrust we have had for the last 10 years," says Houlditch.

In 1988 the average TI employee spent 2.7 days in company sponsored quality-training programs. Each year DS&EG spends the equivalent of 2% of sales, or $44 million, on education and training, including training for quality improvement. Houlditch believes that number is roughly in line with the industry average but adds: "It's how you spend the dollars to get the most output that counts."

In a bid to boost quality awareness among all salaried personnel, in 1986 the group began holding quality improvement workshops in which the principles of the earlier programs were tailored to TI. All 9,000 of the division's salaried workers were slated to go through this program.

Cooney says DS&EG uses its managers as a bridge between external and internal quality-improvement programs. For example, some 80 managers who attended the Crosby quality course returned for a 3-day "train the trainer" course before leading Texas Instruments' in-house quality-improvement workshops.

The DS&EG Training and Education Council, which oversees the quality-training programs, is made up of one representative each from the engineering, man-ufacturing, and quality/reliability operations; a member of the management commit-tee; a human resources development manager; a division or program manager; and the group's personnel manager.

Continuous improvement at TI is aimed at quantifying quality gains. "It's very easy to talk about continuous improvement," says Cooney. "But if you don't have some way to measure yourself, then you really don't know whether or not you're improving." TI measures quality from both ends of its business, from the vendors that supply the company to the customers it supplies. In the case of DS&EG, those customers are either other defense contractors or the U.S. military. For example:

- *Supplier-lot rejects* quantifies the percentage of products from its suppliers that TI rejects for failing to meet its standards. Cooney says this stayed at about 6 to 8% between 1982 and 1990, even though quality standards were steadily tightened.
- *Use-as-is* quantifies the percentage of products the division accepts from suppliers, even though those products in some way do not meet TI's specifi-cations. The products are not defective but may not meet the division's criteria for packaging or documentation or may require some change in manufacturing procedure. This number was steadily reduced, from 2.4% of all supplied products in 1982 to 0.4% in 1989, to 0.3% in 1990.
- The Material Review Board, made up of two TI representatives and a Defense Department official, review reports or actions that result when DS&EG is unable to comply with U.S. military product specifications, manufacturing procedures, or other requirements. The TI group cut the number of such actions from 11.6 per $1 million of billing activity in 1982 to 3.1 in 1989, to 2.5 actions in 1990.
- Quality-deficiency reports are filed by the Defense Plant Representative Office, the in-plant representative of the Defense Logistics Agency (DLA), which monitors defense contractors, when it finds that TI failed to comply with its standards—if, for example, a government inspector discovered a piece of test equipment with an expired calibration-test sticker. The TI division reduced the number of quality-deficiency reports per $1 million of billing activity from 0.44 in 1982 to 0.03 in 1989, and even further to 0.02 reports in 1990.

~ Improving the Product

DS&EG also measures quality advancements through product improvements. Another contractor—which TI declined to identify—once manufactured radar sys-tems for the F-111 fighter that broke down, on the average, every 22 hours of opera-tion. The Pentagon wanted to improve the system's mean time between failures (MTBF) to at least 50 hours and in 1984 gave the contract to TI. The system's MBTF has since been boosted to 125 hours. The HARM missile boasts an MTBF rate of 300 hours, far exceeding the defense department's 125-hour requirement.

TI managers point to other examples of quality payoffs. In 1982, each HARM missile cost about $500,000, but that price tag was cut to around $202,000 for each missile, the result of quality programs along with volume production gains, according to TI. DS&EG even applies a quality yardstick to such nonproduction areas as accounts payable, where improvement is gauged by the decreasing number of errors made in contract-order entry.

The division offers its quality-training programs to its suppliers and even provides its managers to teach the courses. "What we're trying to do is work with our suppliers and make them as good as we are," says Houlditch. About 2,500 vendors supply DS&EG with components that go into its weapons systems. The senior vice president says most suppliers are taking up TI on its offer, while a few are implementing their own quality programs.

Each year DS&EG names about 20 vendors as outstanding suppliers, based on product quality, on-time delivery, and cost criteria. Texas Instuments, however, does not require that its suppliers become Baldrige applicants. "They audit us on a regular basis to make sure we are complying with their quality standards," says Stephen Babler, general manager of the Trident Corporation in Richardson, Texas, which supplies TI with aluminum and other raw materials.

Trident has been doing business with TI for many years, according to Babler, and he observes that in the past 5 years TI has been placing greater emphasis on joint efforts with its suppliers to improve quality. Babler says TI managers periodically visit Trident's facilities to better understand their business, while Babler has attended quality-improvement meetings at TI. Recently he and Trident's chief financial officer attended a TI-sponsored quality control seminar.

Because the Department of Defense places so many restrictions on how contractors buy components and manufacture systems, instituting new programs, even quality improvement programs, can be difficult for contractors, Cooney admits. Recently, however, the Pentagon has been moving to allow some contractors to regulate themselves. The logistics agency has instituted programs to reduce the amount of oversight it has with efficient contractors, allowing it to concentrate regulatory efforts on problem vendors. DS&EG facilities have been highly instrumental in helping to establish the pilot programs.

Between 1988 and 1989 DS&EG's avionics-systems plant in McKinney, Texas, worked with the DLA to establish the criteria and procedures for the in-plant quality evaluation program. The DLA decides which contractors are qualified to monitor themselves based on such criteria as total quality control efforts.

In 1989, the Texas Instruments Lewisville plant was one of four pilot exemplary facilities, identified by the Defense Department that met the procurement agency's principles, demonstrated a quality culture, and manufactured products that met or exceeded department requirements.

∾ Rockwell International: Commitment to Organizational Excellence

"We believe in providing superior value to customers through high quality, technologically advanced, fairly priced products, and customer service, designed to meet customer needs better than all alternatives." So reads a line from Rockwell International Corporation's credo, the one-page document published in December 1987 that declares to the world what the aerospace giant stands for. It is the foundation for the new quality culture, dubbed organizational excellence, that Rockwell is attempting to instill in all corporate employees. So far, so good, but there is still a long way to go.

Both the credo and the organizational excellence program—basically a top-down model for dealing with quality issues—were set in motion by Donald Beall, Rockwell's chairman and chief executive in early 1988. In the mid-1980s, when he was chief operating officer, Beall recognized the increasing competition in both defense and commercial sectors and the limited future of the B-1B bomber program as signs that Rockwell had to bolster its competitive position. The wheels of change were thus set in motion. That meant improving productivity and quality across the company, starting with the worst performers. Beall was fully aware that the initiative had to start from the top.

Rockwell's top management, including the three newly appointed chief operating officers, are all well grounded in organizational excellence. So are the division heads and their staffs. "We are trying as hard as we can to push the concepts of continuous improvement throughout the corporation, both commercial and defense," says Kent Black, one of the three operating officers.

Rockwell is not alone in its quality push. Budget cuts and second sourcing make for a harsher procurement environment for all contractors. As the competition heats up, most are responding in part by improving quality. On top of it all, the Department of Defense has started its own total quality management program and is strongly encouraging contractors to adopt it.

In the mid-1980s, when quality first became an identifiable issue, Rockwell opted for a guinea pig approach. It identified a corporate laggard, changed the top management, and attempted to bolster its operations. Later, when the process of quality improvement was refined and institutionalized, it was molded to fit other divisions. Today, most of the 13 major divisions boast some type of quality program. At the Automotive Division, gain sharing, a type of employee incentive program, is part of the quality initiative. At the North American Aircraft Division, a program called top quality management has been put in place.

At Autonetics Electronics Systems, manufacturer of guidance and control systems for missiles, management combined the Defense Department's quality management initiative with Rockwell's organizational excellence. This marriage of quality principles resulted in product and cross-discipline work teams, statistical process control, and quality function deployment, all of which led to the group's Air Force Logistics Command Blue Ribbon Contracting Award in 1989.

The Rocketdyne Division has a program called commitment to excellence, and since its inception in 1984, employee training has increased 100%. The space shuttle engine maker estimates that its productivity improvement and quality enhancement program saved the company $5.6 million in 1987. The division's engine success rate is 99.4%.

Rockwell executives warn that the investment community should not look for quick financial results from the firm's quality push, especially with the B-1B program winding down. They explain that total quality management is about cultural change. "And you'd better have the long-term commitment and patience to allow the process to work, because you'll never have an overnight success—it takes years," operating executive Black said at a Rockwell management meeting in 1987. The message was clear: Without a quality initiative, Rockwell might not be around to worry about short-term results.

For now, Rockwell points to the qualitative improvements as evidence of its progress: better relationships with its customers, including awards from Canon, Digital Equipment Corporation, and the National Aeronautics and Space Administration; higher quality products from suppliers; and a lower cost of quality due to less scrap and rework.

The humble origins of Rockwell's quality initiative can be traced to the Semiconductor Products Division, today a part of Rockwell Communication Systems and currently a major player in the facsimile modem market. In 1985, the semiconductor division was in a crisis. It was considered the corporate leper in terms of financial performance, recalls Charlie Kovac, vice president for major programs, marketing and international for Rockwell Communication Systems, and part of the turnaround team. The semiconductor operation hadn't been profitable since 1980.

In 1983, Rockwell hired a new president for Rockwell Communication Systems, Gil Amelio, from Fairchild Camera and Instrument Corporation. The following 2 years were a time for internal self-assessment and redirection. On a strategic level, the group decided to target the communications market with a proprietary product. On a tactical level, the company established a management position of director of quality, reporting to Amelio. A series of workshops, the devising of measures of quality, and the revamping of incentive programs followed.

During this period, the "organizational excellence" process began to evolve. Every Friday, the semiconductor group's management met over lunch to discuss quality issues that ranged from mislabeled packing slips to vendor quality. The top 10 problems were studied during the following week and reported on the next Friday. "We didn't eat very much but we sure as hell yelled a lot!" Kovac recalls.

The division set about on the path toward continuous process improvement. Soon good things began to happen. The group turned its first profit of the decade in 1986. Between 1983 and 1988, it doubled its sales, the company claims. By 1990, the chip operation controlled about 65% of the fax modem chip market, much of which is based in Japan, says Roger Steciak, a senior semiconductor analyst at Dataquest.

In 1986, about the time that the semiconductor group was getting its shop in order, the missile systems division in Atlanta, a member of Rockwell's electronics operation and a maker of guidance systems for tactical missiles, was going through its own crisis. It had moved from Ohio in 1981, losing a large percentage of its employees, and was having trouble moving into manufacturing from research and development. In its primary production program, the Hellfire missile, the group lagged on delivery. As with the chip division, the beginning of the transition came with a new general manager, Paul Smith, who moved from a another Autonetics division to head the missile systems division in January 1986.

"The missile operation is in a niche market," says David Jacobs, director of organizational effectiveness. "We realized that in order to be more competitive we were going to have to change." During 1986 and 1987, the company started a number of unrelated quality programs, including statistical process control, just-in-time manufacturing, and gain sharing. At the same time, Smith attended a productivity-through-quality executive seminar at the University of Tennessee, where he met Bill Conway, a quality consultant. By the end of 1987, Smith had brought in house Conway's course, "The Right Way to Manage," and was customizing it to fit the missile group's needs.

By early 1988, all these disparate elements, combined with the organizational excellence process borrowed from the semiconductor division, evolved into what the missile operation today calls its total quality system. It involves educating its 1,800 employees in the use of such specific tools as statistical process control, flow charts, quality function deployment, and Taguchi methods for coupling design and manufacturing, as well as rewarding employees for adherence to quality objectives.

So far, 280 managers and professionals have been through an 8-hour Conway course and nearly 600 employees have taken a 20-hour statistical process control course, Jacobs says. All new hires get some awareness training, too. Similarly, all employees take part in a functional support plan which sets goals for a functional group to improve its performance in one area during the year.

The missile division also restructured incentive systems so that quality improvement and team efforts are paramount. Evaluation of performance is based on six indicators measured monthly: cost of poor quality, hours per unit, average assets, product yield, cost performance, and indirect budget performance. The payoff at the end of the year is based on how well all employees perform on these standards.

The missile systems division's total quality system program goes one step beyond the Defense Department's total quality management program. David Jacobs, the divison's director of organizational effectiveness, defines quality management this way: "To continuously improve our ability to meet customer needs through a systemic approach to managing organizational change."

The quality push produced some marked results. Some include:

- The AGM-130, a powered stand-off missile for the Air Force Tactical Air Command, was threatened with cancellation because of poor test results. It has since been resurrected because of markedly improved quality.
- On one component of the Hellfire, a U.S. Army and Marine Corps antiarmor missile, a cross-functional team of engineers improved yield by 17% over a 4-month period using statistical process control techniques. That saved the division about $200,000 a month in rework.
- Using similar techniques, a cross-disciplinary group managed to cut scrap costs by 87%, reduce rework cost by 80%, and increase the acceptance rate of a component by 98%. The work force was reduced from 20 people to 5 in 8 months.
- In the finance department, reducing the number of late time cards turned in at the end the week enabled the division to save an estimated $70,000 annually. The changes include streamlining the processing of time cards and tracking the results.
- Over an 18 month period, a team of transportation and material services employees reduced dock-to-stock time by 30% and increased accuracy by one-third using statistical methods. They saved more than $300,000 in 1990 and contributed nearly $400,000 to the division's employee gain sharing incentive program.

Beyond all these positive results, the most significant achievement could well be the new relationship the missile division is test piloting with Defense's purchasing arm, the Defense Logistics Agency (DLA). Called the in-plant quality evaluation initiative, the program could mean product inspection will be replaced by a process audit. Rockwell's missile group was the only defense logistics contractor picked out of the Southeast region of the United States to participate in the pilot program. If the initiative proves successful, the outcome could lead to a profound difference in the way the Pentagon does business with its most quality-conscious contractors.

Like the semiconductor products division, the missile division is held up as an example to other Rockwell units; and rightly so. Since 1985, sales per employee have risen by 70%, according to Jacobs. The division also set all-time production records on the Hellfire. How is the success viewed around the rest of the company? "Other divisions at Rockwell come to us," Jacobs comments, as they develop total quality plans of their own.

∾ Other Contractors: Quality Efforts Sweep Through the Defense Industry

Texas Instruments and Rockwell International are not alone in the push toward improving quality. Most other aerospace and defense electronics companies are involved in the process, too—some well on the way and others just getting started. All are prompted by defense budget cuts, by increasing foreign competition, and by the momentum behind the total quality management movement at the Department of Defense.

Following are some examples of innovative quality improvement programs under way in the industry, based on a quality management report published by the Aerospace Industries Association.

- E-Systems, Inc. has a companywide total quality program called the discrepancy data system that attempts to eliminate discrepancies in products and processes. Using statistical process control, it analyzes inspection reports to pinpoint variation, then takes preventive action.
- General Dynamics Corporation's history in quality programs dates back to 1981 at its Fort Worth, Texas, division. The focus of its corporatewide quality push is on measurable goals, management commitment, employee awareness, and training.
- Hughes Aircraft Co., a division of Hughes Electronics Corporation, has a companywide total quality program where each division generates its own list of quality objectives. Performance is evaluated against measurable benchmarks. The program is focused on prevention rather than correction.
- Lockheed Corporation has a policy called star quality devoted to providing customers with products and services that satisfy their expectations of quality. Each division tailors the program to suit its needs and its customers. Concepts include improvement in reducing the cost of quality, employee awareness, and supplier involvement.

- Martin Marietta Corporation uses performance measurement teams. These cross-functional (multidiscipline) groups identify deficient quality, determine solutions, and fix problems. Results are tracked using statistical methods.
- Raytheon Co. has two on-line, real-time monitoring systems for inspection and assurance testing. Both programs monitor data and indicate deviations, so corrective action can be taken.
- TRW, Inc.'s Space and Defense Sector's quality awareness and continuous improvement program has four initiatives: a corrective process for improving design, quality improvement for parts acquisition, upgrading systems and standards through a quality enhancement training program, and the development of quality measurements that tie into the business management system.

Chapter 15 ⌒

Software Developers Seek Better Measurements

American companies have built reputations as the world's premiere developers of software. When it comes to instituting procedures to ensure the quality of their products, however, software developers lag far behind equipment manufacturers. This state of affairs is hardly surprising, given the relative immaturity of the software field. While manufacturers have had centuries to experiment with production, software developers have been forced to transform their operations from an art form into a science in decades.

By their own account, software vendors still have a long way to go. "There are not many successful processes established yet for developing and testing the quality of software," according to David P. Reed, vice president and chief scientist for spreadsheets at Lotus Development Corporation Cambridge, Massachusetts. "The Massachusetts-marketed software industry hasn't found any methods that seem to work too well."

Data collected by the Software Engineering Institute, a Department of Defense–funded research-and-development center located on the Carnegie Mellon Institute campus in Pittsburgh, supports Reed's analysis for software developed in house as well as for commercial packages. Watts Humphrey, director of the institute's software process program, says any company's operations can be ranked according to the following five-level maturity framework scale:

- *Level one.* A company's activities are unplanned and chaotic.
- *Level two.* Operations are somewhat stabilized and an intuitive routine is established.
- *Level three.* The processes in use are defined in a formal way.
- *Level four.* The company measures the processes and begins to collect hard data about their efficiency and usefulness.
- *Level five.* The data collected from the measurements are used to optimize the processes and to progressively improve them.

"We've collected data on several hundred software-development projects in the U.S.," according to Humphrey, "And 75 to 85% are at level one; 15 to 20% are at level two, and very few are operating at any higher level."

To improve quality, software companies should work toward satisfying the Malcolm Baldrige National Quality Award criteria, many experts argue. The Baldrige Award is focused on continuous improvement aimed at achieving the highest level of process maturity, and that sort of steady effort is what's needed to optimize operations, Humphrey says. But due to their lack of process sophistication, few software organizations can yet compete seriously for the award, and even fewer have made the effort.

Mentor Graphics Corporation, the Beaverton, Oregon, developer of electronic design automation (EDA) systems, is one of the exceptions. "We made the decision to compete shortly after the Baldrige Award was established," says Gerard H. Langeler, who was president and chief operating officer at that time. The company applied for the first time in 1990, but it was cut from the competition in one of the preliminary rounds of evaluation, he reports.

Mentor went through a major overhaul of its development methods in 1987 and has since integrated the Baldrige criteria into its current development processes. For instance, Mentor uses "cross teams" made up of software engineers, customer-support and marketing staffers, and even key customers in the development of its software. Involving customers in the process helps to shorten product-development cycles and to clarify customer requirements.

Software vendors are motivated to produce high-quality products for the same reasons that are hardware manufacturers: competitive advantage, reduced service and upgrade costs, and improved customer relations. The challenges faced by programmers, however, are often unique.

One of the most obvious of these challenges is the rapid growth in the size of commercial programs. As the number of lines of code increases with each successive release of a package, so does the possibility of error. At the same time, the days when a single creative programmer could produce an innovative and successful program are largely past. Today, tens and even hundreds of programmers may contribute to the development and testing of a single product, and each of them could introduce errors.

Different versions of the Lotus 1-2-3 spreadsheet, for example, contain from 250,000 to 330,000 lines of code, says Reed. From 5 to 10 software engineers develop the basic architecture for such programs, 30 to 50 programmers write most of the code, and another 30 to 50 engineers are involved in testing the program, he estimates.

Many software systems are even more massive and complex. Mentor Graphics' high-end EDA software package, for example, contains more than 10 million lines of code, according to Langeler. Projects of this scope require teams of programmers to work on separate modules of the program that must later be merged seamlessly. The length of time of the job adds to its difficulty. A typical software program takes about 2 years to move from the concept stage to the market. "The competitive environment can change often during that time, and modifying the software to reflect those changes can be difficult," according to Reed.

Many software-development methodologies require that developers freeze the specifications of the program up front. For Massachusettsive programs requiring many months or even years of development work, however, Reed explains that "it's not feasible to set the requirements for the program early and then not allow changes later."

Compounding the problem is the need to regularly upgrade programs already in the field—a process that can introduce new errors along with the new features. "Most of our products have lifetimes of about 4 to 5 years but get significant enhancements at least every year," says Dennis R. Schnabel, director of corporate quality at Mentor Graphics.

The software industry's key challenge is a very basic one: how to objectively measure the quality of a program. After all, a complex program is deployed differently by every user, and traditional hardware parameters such as mean-time-between-failure don't apply. Quality measurement for software has always been some variation of the number of bugs per 1,000 lines of code. But that measure often has little bearing on the perceived quality of a program. "For some programs, one bug per line isn't bad; for others, 0.01 bugs per 1,000 lines is lousy," says Boris Beizer, a consultant in software testing. While all developers attempt to reduce the number of bugs in their programs, the process can be haphazard. In tests of defect detection and removal efficiencies, most developers find fewer than one bug in three, says Capers Jones, chairman of Software Productivity Research, Inc., a consulting firm in Burlington, Massachusetts.

The low efficiency of bug detection is due in part to the growing complexity of software programs themselves, says consultant Beizer. "Advances have improved programmers' productivity, but they haven't reduced the total bugs-per-source-code statement," he remarks. "As the complexity increases, the bugs get nastier, subtler, and tougher to find."

In any case, measuring the quality of software by its bug count alone misses much of the picture, points out Langeler at Mentor Graphics. To better assess overall quality, the company has adopted a system called FURPS (functionality, usability, reliability, performance, and support) developed by the Institute of Electrical and Electronics Engineers in the early 80s. Evaluating software in all the FURPS categories gives a better balanced view of quality than reliability measurements alone. "A program might have zero bugs, but if it doesn't provide much value for the user, it's not high-quality software," Langeler explains.

While FURPS defines important software attributes, the way to implement and measure them is left up to the developers themselves, according to Mentor Graphics' Schnabel. "We took the general FURPS categories, determined which attributes the categories were trying to measure, and developed a system to measure them." For example, the measure of a package's functionality includes both the percentage of functional specifications reviewed and approved and the percentage of enhancement requests from customers that have been implemented in upgrades.

While the FURPS concept is useful in developing quality software, it isn't extensive enough, says consultant Beizer. His clients complete a questionnaire containing about 500 queries that examine software development and testing processes in much more detail than FURPS, he claims.

While there are no hard-and-fast answers to the software quality question, developers are investing rather heavily to make the production of software more a science than an art. Novell, Inc., for example, employs one test engineer for every two development engineers, according to Richard King, vice president of software development. A new program typically undergoes from 1 to 4 months of system testing and then several months of beta testing at selected user sites, King explains. Programs usually get two or three updates during the beta phase, a period that "gives you a last vote of confidence about your software," he says.

Software firms are also turning to new programming languages and techniques to improve both productivity and product quality. Mentor Graphics, among others, relies more on object-oriented programming, which helps developers break large programs into smaller modules with data shielded from users and procedures built in. The object-oriented approach also enables programmers to more easily reuse code in future packages. By using the inheritance capabilities of the object approach, once generalized objects have been developed, new functions can be added by constructing lower-level objects that draw on the properties already coded and tested in the generic objects.

Object-oriented programming is not a panacea, however, warns Beizer. "It is much harder to test than ordinary software, and it could cost four to five times as much to design initially," he says. Testing is more difficult, he explains, because object-oriented programming tends to hide the complexity of the code and obscure any bugs that are present.

Beyond programming languages, various tools can automate testing procedures and track and record bugs. The tools are not foolproof, however, and there are limits in their capabilities to find all kinds of potential errors, unforeseen interactions, and weaknesses. Thus, most developers stress that they still must focus their efforts on understanding the underlying processes that can assist in quality assurance and testing. Further complicating the software quality problem is that in different application spheres the perception of what constitutes quality may vary widely.

∾ The Quality Struggle in Design Automation Software

A good example of the varying perceptions of quality can be found in the electronic design automation business. For developers of EDA software, life with the quality issue has not been easy. Design automation vendors are faced with trying to measure a "manufacturing" process that is performed by highly creative individuals or groups with no production lines to sample. When the "products" reach the market, customers consider response time and service major factors in evaluating an EDA vendor. This puts a different spin on quality procedures than is traditionally seen elsewhere in the electronics industry.

Hardware manufacturers, by contrast, have been able to establish definitive quality assurance programs. The formula for these manufacturers has been easy to define: either the product works or it doesn't, it meets specifications or it doesn't, it serves the customer or it doesn't. For hardware producers, the metrics are there. Elementary math tells them that parts-per-million failure rates are more desirable than parts-per-hundred. Statistical process control is straightforward when there is a steady process to measure.

The definition of software quality emerging in the EDA world sounds simple enough. It's that the software does what the customer wants when the customer needs it done. But what if every customer wants software tuned to its own development methods, workstations, product line, and timetable? What sounded so simple can turn into a nightmare. While attempting to satisfy customers, the EDA vendor also must implement some kind of process control for its internal development and manufacturing cycle in order to track performance. Further, time must be available for creative engineers to develop new products and to expand existing lines. Keeping tabs on quality without causing a drag on this creative effort can be a delicate balancing act.

Several EDA vendors say they already have a handle on quality. Others have identified it as the issue for the 1990s and are striving to come up to speed fast. How well the quality issue, in all its pervasiveness, is handled could play a large role in each vendor's continued success. "If you don't start a company with the right basic beliefs then you have a difficult time recovering. We tried from day one to establish cornerstones that incorporated quality," says James T. Hammock, president and chief executive of EDA vendor Silicon Compiler Systems Corporation (now a subsidiary of Mentor Graphics).

The cost of quality is very real to the EDA industry. Using engineers' time as a measuring rod, it is easy to put a dollar value on quality, says Lewis White, senior vice president for software engineering for Daisy/Cadnetix, Inc. of Mountain View, California. The average development engineer costs $100,000 in compensation and is expected to produce $1 million in revenue per year, White says. If 25% of an engineer's time is used to solve customer problems that stem from a lack of quality, it costs the EDA vendor $25,000 in the engineer's time, killing $250,000 in revenue.

White says another value of quality is less tangible but just as real. "It is just a matter of time before quality becomes a marketing edge," he says. As with the electronics industry as a whole, EDA vendors face intense time-to-market pressures. Customers expect updates and new products more frequently than in traditional industries. But as design automation products become more complex, this becomes harder to do.

"Release cycles are taking longer. EDA vendors have to shorten the cycle time to take care of hard bugs and enhancements. There have to be a minimum of two releases a year," says James Hogan, director of corporate design automation for National Semiconductor Corporation.

Customers are looking at the total picture when defining the quality of their EDA products. They want software that works the first time, is delivered on time, and is well supported and updated in a timely manner. "It does not bother us to find bugs," says Patrick Scaglia, vice president for product development for supercomputer-maker Evans & Sutherland Corporation, "What is important is that they (the vendors) start to work on the problem immediately."

⌒ *Partnering Helps Keep Software at the Cutting Edge*

One key to quality programs at Silicon Compiler Systems and other vendors is customer involvement in product development. To get the input it needs when developing new products, Silicon Compiler Systems works with early adapters, companies such as Motorola and Sierra Semiconductor, that are constantly pushing the limits of existing EDA tools and demanding new ones. These partners receive early code to help them get an edge on developing their new products. In return, they send feedback to Silicon Compiler Systems that helps Silicon improve the software.

Silicon Compiler Systems also goes out of its way to make its employees responsible for and aware of quality issues. The central tenet of its philosophy is "you eat what you cook" says Hammock. The software development group is responsible for maintaining the product after it is in the customer's hands.

"If they push the functionality at the expense of quality they have to fix it," says Hammock. The San Jose company tracks bugs back to the developer who created them and that engineer then "owns" that bug. The bug is his or her responsibility to fix; points are assigned for each bug he or she owns, and bonuses may be on the line.

The company also employs such methods as regression testing that are common throughout the EDA industry. These methods ensure that new software works with existing software and retains all the features of previous releases. A key to quality software is making upgrading painless for the customer.

Mentor Graphics, the industry leader, has established a strict regimen to ensure, track, and maintain the quality of its products. Applying for the Baldrige Award was an important part of this process. "It forced us to write down what we were doing," says Mentor's Langeler. "We got to a few places where we drew blanks. There were a lot of things we weren't measuring." Mentor aims to win the award by 1994. Even though "winning the Baldrige Award is the goal, the journey is more important," says Langeler.

Mentor has, however, already achieved another quality goal by becoming a certified Class A MRP II manufacturer. Certification comes from an outside group of consultants, Oliver Wight Co. of Essex Junction, Vermont. The manufacturing resources planning process offers companies a standard by which to judge the effectiveness of their manufacturing efforts. From the beginning of 1985 until January 1988, Mentor worked to earn the the Class A MRP II ranking, for which, says Langeler, only 5% of U.S. companies qualify. To qualify, a vendor must answer yes to 23 of 25 questions on how it manufactures product, gauging such areas as scheduling, inventory control, planning, and education of workers.

In addition to these external measures of quality, Mentor also applies strict internal guidelines for each new release. New releases must have fewer bugs, higher test coverage, and solve any problems a customer may have discovered. "If the product has not met the guidelines it is not released. No one person can say yes to a product shipping, but a lot of people can say no," says Mentor's director of corporate quality, Dennis Schnabel.

Companies such as Mentor/Silicon Compiler Systems, and Daisy/Cadnetix are honing quality assurance programs into marketing weapons and using these same weapons to cut down product development cycles and increase the efficiency of their organizations. Meanwhile, other software firms are struggling to develop the proper mind-set and approaches to get total quality programs rolling. At Cadence Design Systems in San Jose, California, Bruce Bourbon, vice president for marketing, explains, "We don't want to impose panaceas for development (using strict methodologies). We are trying to walk a fine line of not stifling creativity."

Cadence developed a quality assurance program that spans 45 major products that are in various stages of maturity. Part of Cadence's system requires that the engineer who writes the source code also provides test for the code. This effort is to make designers aware of quality issues from the beginning. Working in Cadence's favor is its rapid growth, which necessitates the frequent hiring of new engineers. "Every engineer has a 2-day training class to introduce standards, our framework, and the engineering guide," says Michael Macfarlane, Cadence's group director for software methodology.

Quality is an especially important issue for another industry heavy hitter, Valid Logic Systems Inc. of San Jose. Valid is trying to integrate a broad line of tools from several companies and several different development methodologies. "The quality issue is driving the market because it has not been met in the past. In the past, the market has grown so fast that customers have not pushed very hard for quality tools," says Frank Wypychowski, Valid's vice president for engineering in the CAE (computer-aided engineering) division.

⟳ IBM Strives for Error-free Shuttle Software

While much of the software industry struggles to implement rudimentary quality-assurance procedures, a team within IBM's Federal Sector Division has established an operation that generates virtually fault-free code. The Houston-based team had little choice in the matter: It writes the on-board software for the space shuttle. "Our contract with NASA says the software will be error free," says Ted Keller, senior systems engineer for on-board software systems in the space-shuttle program. "We're producing *man-rated* software—that on which human life depends."

Keller admits that no software of any complexity can be proven to be completely without bugs, but a series of steps implemented by the Houston team has dramatically cut the number of errors in the shuttle software. Once IBM has tested and certified the software for flight readiness, that number now stands at about 0.1 to 0.2 errors per 1,000 lines of code. By industry standards, that error rate is very good. Based on the five-step maturity rating for software development—devised by the Software Engineering Institute, a software research-and-development center funded by the Department of Defense—NASA evaluators noted that IBM's shuttle software development program is a rare one that has reached the highest rating.

The first stage in the development process is a requirements-evaluation process, Keller says. A group of requirements analysts ensure that the functionality of the software is clearly stated and understood and helps translate these requirements for the programmers who must write the actual code. "We reduced the total number of errors by 20% by implementing this step alone," Keller says.

Next comes the software-development phase, which includes formal design reviews and formal code reviews as the program is written. Once written, software modules are tested separately to ensure that they perform their function properly. They are then tested in combination to check the performance of the full system. The system test includes running the software on exact replicas of the shuttle's on-board computers and simulating every phase of a mission, from launch to landing.

Finally, the Houston team turns the software over to an independent verification and validation (IVV) group within IBM. This IVV group participates in the initial requirements-evaluation process. Then, while the software is being written, the IVV staff builds a set of test cases to put the completed programs through their paces. The Houston developers and the IVV testers have "a benevolent adversarial relationship," says Keller. "They compete to see who finds the errors."

Despite the varying stages of implementation and maturity of quality programs across the software world, certain facts ring true in all cases. Quality and the quest for it must come from the top down. It is a cultural issue as much as an issue of measurement and methodology. The other universal principle is that the quest for quality is a road with no destination. Once a quality goal is reached, another must be set. The journey is more important than the destination.

Part III ∼
The Total
Quality Approach

Chapter 16 ∽
Satisfied Customers
Survival's Bottom Line

United States companies whose customers are other businesses used to have it easy. They would just come up with an innovative product or service, get potential customers to agree to some acceptable quality level (AQL), and then, if things weren't too busy at the time, ship within a few weeks or so of the promised delivery date. Any firm with such an approach to business, if it still survives, has had a rude awakening of late. More and more large companies, the big customers for business-to-business products and services, have put a much louder snap into their whip-cracking. In those good old days, a flub-up here or there might mean a reduced portion of the business, but a timely price cut could win back the lost ground later.

Now more and more of these big customers are looking far beyond the prices they are being charged. They are radically trimming vendor lists and demanding to know a lot more about each supplier's operations. These big firms are being pressed not just by tough global competitors but by other U.S. firms that have turned to the total quality approach. They recognize that to thrive, and maybe even to survive, they must bring chosen suppliers into their quest for better performance and higher quality.

When Anthony F. Lefkowicz visited Ford Motor Co.'s electronics unit in 1988, for example, he learned that his company, Augat, Inc., had won the dubious achievement of being named Ford's worst supplier. At the time, Augat was banned from applying for new business from Ford, the source of 40% of its $103 million automotive-electronics business, until it cleaned up its act. "We were in a death spiral," according to Lefkowicz, vice president and general manager of the automotive division of Augat, a $300 million connector company located in Mansfield, Massachusetts

Augat saved itself by heeding Ford's increasingly aggressive call for quality. Since adopting Ford's quality survey and quality-operating system, Augat has invested more than $1 million in tooling, process control, and training in a concerted effort to win preferred-supplier status under Ford's Q1 certification program. "We adopted every quality recommendation we could steal from them," says Lefkowicz.

The result was that all five of Augat's automotive operations achieved Q1 status and the company won $15 million worth of new business with Ford during

1988. Even General Motors, which has long made its own connectors, has expressed interest.

Alas, such success stories are all too rare. With U.S. companies purging their vendor lists of marginal suppliers in favor of long-term partners, those unwilling to invest in quality can expect to pay a stiff price in lost market share and profits. Xerox Corporation, a winner of the Malcolm Baldrige National Quality Award, has slashed its supplier base from 3,000 to 350 since 1981. IBM cut its roster of significant suppliers from 4,000 to 3,000 in 2 years. Unisys Corporation from 2,300 suppliers in 1985 to 670, and Motorola, Inc.'s communications division in Schaumburg, Illinois, has trimmed down from 4,000 to 1,500 in the same period. "We'd like to have as many single-source suppliers as we can," according to Motorola's Ron Vocalino, regional sourcing manager for the division. "We'd like to consider the supplier an extension of our own factory."

Judging by the number of casualties, few suppliers are meeting such tough criteria. "The suppliers haven't been able to keep up with us," says Motorola's Vocalino. "It's the quality of their parts that is keeping us from reaching six-sigma," an exceedingly tough quality standard that calls for no more than 3.4 defects per million operations.

Motorola is not alone in its frustrations. More than half of the 262 senior managers who responded to a recent survey by the American Electronics Association (AEA) cited vendor quality as a primary obstacle to profitability. The AEA survey portrays an industry that is aware of but uncommitted to the need for total quality:

- About 90% of those polled named quality and reliability as one of their top three concerns, but only 22% have begun their own total quality management (TQM) programs.
- While 93% of the respondents said that time-to-market is important to their business, only 5% had adopted systems for shortening their product-development cycles.
- While 46% named timely delivery as a top priority, only 16% had implemented just-in-time (JIT) systems.
- Only 15% had adopted electronic data interchange (EDI), by which business transactions are handled electronically rather than on paper.

To close these performance gaps, progressive companies are dragging their suppliers into the quality arena. Companies including Hewlett-Packard, IBM, Motorola, and Texas Instruments offer their in-house quality-training programs to suppliers. Motorola itself has given basic quality training to 7,000 companies, and company executives make more than 500 quality-related speeches a year, notes Richard Buetow, vice president and director of quality for Motorola. Winston Chen, president of San Jose-based contract manufacturer Solectron Corporation, another Baldrige winner, says he attends at least one customer-sponsored quality day per month in order to stay on top of customers' quality requirements.

Measuring up to the expectations of the quality crusaders is an expensive proposition, especially for midsize and smaller suppliers. According to the AEA survey, the average TQM program takes 13 worker-years and $468,000 to implement. JIT takes 9 worker-years and $300,000, and EDI takes 3 worker-years and $286,000. Quality training itself costs an average of $127,000 a year for 60 companies polled by *Electronic Business* magazine, with one company spending $1 million.

While those up-front investments may be painful, observers say that those who fail to make the transition to greater quality are bound for corporate ruin. "I don't know if we're putting people out of business," says Motorola's Vocalino. "But we are certainly upping the ante."

∼ A Tougher Task for the 1990s

Not all suppliers are waiting to be hauled, kicking and screaming, into quality. Many are moving on their own to learn more about their customers—who they really are, what they demand, and what the vendor needs to know to consistently meet those demands.

Suppliers aren't alone in their perceptions of the importance of the customer in this era of total quality. In scoring applicants for the Malcolm Baldrige National Quality Award, the judges give heavy emphasis to customer satisfaction. Offering even more proof of industry's awareness of the buyer, specialized consultants report brisk new business from companies seeking to polish their images among customers. "We're having our best year ever," beams Walter P. Smith, president of Prognostics, Inc. in Menlo Park, California, a consultant in customer surveys for high-tech companies. Where executives once tended to discount the value of such surveys, he says, "Now they know that they need help to get a competitive edge."

Even that recognition by itself doesn't go very far in luring new customers, or holding on to established ones, when buyers' needs and expectations change almost daily. Many executives complain that customers often can't even articulate their needs; they know only that they have problems and expect the supplier to solve them. When a customer of Shipley Co., a Newton, Massachusetts, supplier of photoresists and specialty chemical systems, complained that its new automated mail-handling equipment couldn't deal with packages from Shipley, Patricia J. Adamcek, Shipley's marketing and planning manager, says Shipley designed special packaging for that one customer.

For most companies, the task of keeping customers happy will get tougher during the 1990s. For one thing, the growing complexity of the marketing/distribution chain makes it difficult for manufacturers to identify their customers or learn how satisfied they are with the product. "Is our customer the purchasing manager? The manufacturing person? The person who buys the car?" asks Peter Gedvilas, director of quality assurance at Cherry Electrical Products in Waukegan, Illinois, a $214 million producer of switches, semiconductors, and other electrical products, some of which become part of automobiles. "All of these are users, but they all have different needs," says Gedvilas. To learn more about problems at each stage in its automotive electronics market, Cherry personnel regularly visit auto plants and car dealerships to learn how often Cherry components must be serviced.

Another problem is that many companies assume that they are meeting all the customers' objectives simply because they provide quality products. "You can meet the specs and still not help the customer solve his problem," explains Gedvilas. What is needed today, he says, are total solutions based on quality goods and services. No longer do auto engineers try to become experts in switch design, for example. "Now they just give us the overall size and performance standards," says Gedvilas. "The rest is up to us."

More and more, it falls to the supplier not only to fill the buyers' needs but to predict those needs, perhaps even helping to create them. Few consumers of the 1950s realized they needed instant photos, for example, until Polaroid Corporation created the need with the Land camera. Sony later did the same thing with its wildly successful Walkman. If such marketing phenomena are to occur regularly, however, industry must adopt "new quality methodology by which someone detects the early symptoms of customer demand and provides solutions," explains Shoji Shiba, visiting professor at MIT's Sloan School of Management.

～ Looking for Bad News

While few innovative companies strike it quite as rich as Polaroid or Sony, suppliers have recently turned to various methods to help them detect the early symptoms of customer demand. Dover Electronics Co. in Conklin, New York, for example, began sending letters and two-page surveys to as many as 200 of its 1500 original equipment manufacturer (OEM) customers every month, according to president Ronald R. Budacz. "We ask them whether we're meeting their needs and how we can do better," says Budacz. Following up each letter with a phone call, he says, generates response rates of more than 95%. "If we get negative comments, I call for more information." When customers reported problems in reaching Dover by phone, for example, the company made it policy that all calls are answered by the third ring.

Convex Computer Corporation, a producer of mini-supercomputers based in Richardson, Texas, uses questionnaires for semiannual customer-satisfaction audits it has conducted since 1985, according to Terry Rock, senior vice president of operations. Convex also measures customer satisfaction via independent research firms such as Dataquest and Prognostics. Shipley researchers also are developing a questionnaire to track customer satisfaction across several purchasing criteria. "We need to know where we, as well as our competitors, stand," says Shipley's Adamcek. "That tells us what we have to shoot for."

Such methods not only keep suppliers tuned in to customers' needs and complaints but provide surprising insights. "We used to think that price was the most important parameter for our customers," says John C. Kohler, vice president of product support services at Tellabs Inc., a Lisle, Illinois, producer of telecom equipment. But a customer survey conducted as part of Tellabs's strategic quality planning revealed that price was far less important to buyers than product reliability, on-time delivery, and responsiveness.

Gathering information from buyers must go far beyond routine surveys and questionnaires, however. Just as important, says Convex's Rock, "You have to talk to the customer often to learn how he's going to use this technology. If he needs different third-party software, for example, we try to line him up with the right vendor." On those occasions when things go wrong, he adds, top executives must be prepared to drop everything and "get their butts out there."

~ *A Common Language*

How many different quality programs can any one supplier satisfy? While larger companies can keep pace with most of their customers' quality demands, smaller companies have to struggle to keep up. "The biggest problem is that we are asked to know so many different quality languages," says Mike Werner, a vice president of manufacturing at semiconductor fabrication equipment-maker GCA Corporation "It would make life easier if there were some common way of communicating. As it is, the cost to the industry is not trivial." Just in the semiconductor arena a number of standards organizations, including the Electronics Industries Association, based in Washington, D.C., the International Standardization Organization, and the Austin-based semiconductor consortium Sematech, have all developed quality standards. None has been widely accepted as of yet.

In the absence of a standard, a variety of company-specific supplier evaluation programs and product certification programs have sprung up. Many companies have instituted self-certification programs through which vendors can save their customers time and money in incoming inspection costs by guaranteeing their ability to consistently meet quality requirements for a given product. For example, industrial-control maker Allen-Bradley, a Milwaukee-based subsidiary of Rockwell International Corporation, has reduced the amount of rework it does on vendor-supplied components by 60% and cut vendor-supplied rejects by 92% since 1985 through such a program.

Increasingly, however, suppliers are being evaluated not only for the quality of their products but for the quality of their entire operation. Solectron's Chen says that applications for evaluation programs frequently require financial statements along with statistical process control (SPC) data. "In the past, the major original-equipment manufacturers in this country always took their suppliers for granted," he says. "Now they've realized that half of their total costs can be tied up in their suppliers—if a supplier isn't financially stable, it could mean big trouble."

Some customers have taken an especially tough line on supplier relations. Both Motorola and IBM have demanded that their suppliers apply for the Baldrige Award as a means to identify areas for improvement. (One division of IBM requires the implementation of a six-sigma program as well.) Other companies insist that suppliers implement SPC, EDI (electronic data interchange), or specific quality-related systems. Just one year after Motorola's request, about 35 of its suppliers had applied for the Baldrige, says Vocalino—more than a third of 1990's 97 award applicants. Some 200 others lost their business with the company altogether for not responding. To be fair to its suppliers, Motorola set up a committee to evaluate the responses of vendors planning to enter the Baldrige competition some time in the future, or in some cases to exempt firms that might be too tiny or in tangential business areas.

Others take a less insistent approach to bringing suppliers up to speed. "Insist is a tough word," says Jim Kemp, director of materials management of Allen-Bradley's industrial control group. "Lots of companies insist that their suppliers use SPC, for example, but there are also a lot of suppliers with file cabinets full of charts and graphs that no one ever uses. All we ask initially is that management of the supplier be willing to commit to a continuous improvement program." Nonetheless, Allen-Bradley cut its supplier base by 25% over a 5-year period.

Buyers, meanwhile, are working to become better customers. Many use rolling forecasts to keep suppliers apprised of their purchasing plans 9 to 12 months in advance and often will share the financial burden of forecast errors or market fluctuations. Solectron's Chen, who was forced to lay off 100 people in 1985 when a major customer reneged on half a big order, says that more customers are now willing to accept liability for inventory already in transit or unneeded materials that had been ordered far in advance.

Even in times of overall industry sluggishness, a growing number of suppliers are making strides to improve quality. Almost half of the 262 respondents to the AEA survey had a company quality policy, and 42% had a committee to oversee implementation. Chen sees an overall improvement in quality in the electronics industry. "Defect levels have gone down, and the definition of quality is expanding," he says. Lateness, for example, is now considered a defect. "It is getting more difficult to differentiate ourselves on the basis of quality."

Other companies are responding in ways that sometimes seem mundane but are important to their customers. Shipley recently set up a group to improve communications with customers who request more user friendly engineering data, according to Gene Goebel, corporate vice president of sales and marketing. As part of its formal quality-monitoring policy, Eaton Corporation in Milwaukee, a producer of programmable controllers among other things, has developed a simulation program to help customers understand the system before delivery, explains Jerald J. Theder, general manager of the logic control division.

Such seemingly simple processes pay off. Shipley's Adamcek notes that attention to buyer satisfaction has contributed heavily to the company's 15% annual growth rate and also has garnered the company a raft of customer awards (including Texas Instruments Supplier of Excellence Award 5 years running). Eaton's focus on quality and customer service resulted in an extensive redesign of production processes; where lead times were once measured in months, according to Theder, about 95% of the products now are shipped within 72 hours.

From 1986 to 1989, Tellabs boosted its overall supplier ratings with one major customer from 84% to 87.5% and is shooting for 90%. Kohler notes that only five of the customer's suppliers are rated at 90% or better. Convex's Rock credits the company's focus on quality and service for its annual sales increases of 50% during the period 1985–1990 except for the second half of 1987.

A result of the quality crusade, say industry leaders, is that suppliers who do not cement themselves into long-term, quality-driven partnerships will lose out as business customers consolidate their vendor lists. Established leaders are likely to gain the most. "I think it's great," says Gerard H. Langeler, former president of EDA (electronic design automation) leader Mentor Graphics Corporation of Beaverton, Oregon. "Demand for quality will dramatically reduce the number who can confidently compete. It's a healthy process."

Midsize suppliers may stand to lose the most. Not only can they not match the up-front investment in quality of larger competitors, but they lack the flexibility of smaller competitors.

∿ Small Firms Grapple with Quality Demands

Smaller companies face a struggle to keep up with the toughening demands of their business customers. Quality-conscious giants, some of their biggest customers, are pruning supplier ranks, foreign players are driving down prices, and bigger U.S. competitors have far more cash to invest for consultants and training and to realign production and other operations.

At the same time, smaller companies do have some advantages. For one, internal and external communications are often easier and quicker than in large firms. For another, small companies tend to be more flexible in satisfying potential customers, particularly in the design cycle. "A small company can achieve this simply through better communication," according to Charles May, quality assurance manager at Appliance Control Technology, Inc. (ACT) in Addison, Illinois, a $19 million-a-year supplier to the $2 billion appliance-control market.

Meeting customers' needs may not be quite that simple, however Training is important so that employees learn to focus on customer needs and to meet quality requirements. That takes time and money. An *Electronic Business* magazine poll shows that small companies with annual sales of $20 million or less spent an average of $24,000 in 1989 on quality training and education. One such company spent a total of $500,000. ACT, an unsuccessful contender for the 1990 Baldrige Award, spent $80,000 on quality training for its 150 employees in 1989, with another $80,000 to $100,000 slated for 1990.

To get around these high costs, many small suppliers rely on their customers' resources. One example is Dynacircuits Inc. in Franklin Park, Illinois, a $20-million-a-year maker of single-sided printed circuit boards. With the help of its customer Ford Motor Co., Dynacircuits instituted a statistical process control program, held 8-hour employee-training sessions, and designed its quality-operating system. Dynacircuits recently won Ford's Q1 Preferred Quality Supplier status. "We saw an advantage in the resources of our customers," according to Randall F. Ferman, president of Dynacircuits.

Still, Dynacircuits spent $400,000 in 1989, or 2% of sales, on quality education. The company also devoted $10,000 to supplier training and to starting an SPC program at one of its major suppliers. Does such training pay off? Dynacircuits reports it has absorbed the additional training expenses with no price increase to customers and no decline in profits.

Training is central to quality efforts at Electronic Space Systems Corporation (Essco) in Concord, Massachusetts, a $16 million defense contractor with 120 employees in the United States and 40 in Ireland. "The problem is the same in big and small companies: finding the right people for the right job," says Stephen M. Quinn, quality-assurance manager. "We've been tailoring our resources and requirements toward quality," says Quinn. "We made the commitment as a company to put the resources in place. In terms of staff, we kept the number the same but improved their capability and flexibility and beefed up the caliber of the people." Essco spent a total of $6,000 over a 2-year period to train its employees in the manufacture and assembly of radomes, the covers for radar equipment.

Essco's commitment to quality extends to its willingness to lose out on jobs by not being the lowest-cost bidder. "We use higher-grade materials," says Susan Cuevas, vice president of marketing. "It has served us well," she reports, by building the company's reputation and bringing in repeat business.

Not many companies are willing to sacrifice business opportunities to maintain quality manufacturing. One way to keep from losing out is to do a good job of communicating the quality elements that might add some costs. At Praegitzer Industries, a $30 million printed-circuit-board manufacturer in Dallas, Oregon, quality assurance director Greg Blount says, "We do everything from a realistic standpoint so we know how much a problem costs and we can document it."

At the same time, wise investments in quality programs can pay off handsomely later. Since launching its quality strategy, says Blount, Praegitzer has reduced cycle time by 15% and improved employee productivity by 10%. Blount estimates the company invested $250,000 in quality education and training programs in 1990 for employees and suppliers. The company has used the Baldrige Award criteria as a benchmark and plans to compete in 1992.

Chapter 17 ∾

Education

At the Heart of Quality

There's a big difference between training and education, especially when it comes to quality. For years, U.S. industry has been doing a reasonable job of training its workers, especially in larger firms with excellent training programs. But now, with international competition as stiff as it is, training alone, even if it's done well, just doesn't cut it. While training teaches skills, education provides the foundation essential for a complete cultural shift. To become stronger global competitors, many U.S. companies are in desperate need of a new culture.

Many companies claim to educate their employees. Indeed, industry now rivals the nation's school systems in the scope of its efforts to educate Americans. Some $210 billion is spent annually on formal and informal employee-based training in the United States, nearly as much as the $238 billion total spent on U.S. elementary, secondary, and higher education combined, according to the American Society for Training and Development.

Some notable leaders exist in the business community, such as Motorola, which spends more than $50 million a year on education and training. While much of that is for skills training—anything from basic English to advanced statistical methods—a good percentage is devoted to courses in the corporate culture of total customer satisfaction. All levels within Motorola, from top management down, receive a grounding in the culture of total quality.

Other U.S. corporations have made a similar commitment to a total quality culture. Some are in traditional industries, such as Milliken in textiles and Ford in automobiles. Wallace Co., a Texas-based distributor/service organization for the petroleum industry which won the Baldrige Award in 1990, has only 280 associates (employees) yet logged more than 19,000 hours of quality training over a 5-year period. The trend is even stronger in the recently beleaguered high tech sector, where U.S. firms were once virtually unchallenged global leaders. High tech companies with operations in Japan, such as IBM, Hewlett-Packard, Texas Instruments, and Xerox have led the push toward total quality. The majority of companies across all industries, however, have yet to begin the journey.

Starting out on a process of continuous quality improvement is a daunting task, not at all like the quick-fix management fads that have come and gone so frequently in American business in the past. It is arduous and unglamorous, requiring extensive communication to all employees and a work force dedicated to achieving tough goals. In this time of celebrity CEOs and high-rolling managers the methodical manager is seldom in the limelight, but it is increasingly clear that U.S. industry needs the measured and studied practitioners more than it needs the addicts of instant gratification, especially when it comes to quality. A new type of enlightened leadership, with a long-range outlook and an abiding respect for the work force at every level, is an essential start toward the shift to a total quality culture.

That doesn't mean that charismatic leadership is no longer vital. Total quality must start with a commitment from the top, and it takes dynamic, respected leaders to get the ball rolling. It is not enough, however, for upper managers to call for a cultural change toward total quality. Just as important, they must practice it. Many toes will be treaded on and mistakes made as workers become empowered and middle management tries to adapt to shifting roles. Top management must provide a steady, supporting hand as the struggle progresses.

Once top management decides to start the quality journey, it must find some way to propagate the message across the total organization. The problem of getting started is a tough one. Management must follow up pronouncements with clear direction. But where can U.S. companies look for the required education?

∿ Look Inside the Company First

According to quality experts, executives must first look inside their own companies. It's important to get support for the total quality effort from the upper management team and then to enlist their help in expanding the program across the whole organization. Books and articles by quality experts, such as W. Edwards Deming, Joseph J. Juran, Genichi Taguchi, and Kaoru Ishikawa, are good sources of inspiration. For a jump start, however, it can be helpful to get some hands-on advice. Getting outside help may seem easy because there are so many consultants pitching their wares and institutions offering courses in the quality field, but the very diversity of options can make the choice of where to start overwhelming.

Again, the answer lies within. Gaining a sense of direction can help immeasurably in evaluating potential outside gurus. "I'd take a penetrating look at what I am trying to accomplish in the company first," recommends Mark Parrish, former president of Datachecker Systems, Inc., based in Santa Clara. In 1987, the company, then a division of National Semiconductor, began its quality revolution. Parrish's opinion is echoed by the majority of experienced executives and consultants. Without a clearly defined mission, they say, there is no way to understand why the company is in need of a change, what benefits might be achieved as a result, and which experts seem to be on the same wavelength.

In Parrish's case, as with the majority of American businesses that set out to change their culture, the introspection was a result of crisis. After 14 successful years making laser-based grocery checkout systems, the company lost its ability to innovate quickly enough to keep up with the market. "It became apparent that our culture had accepted mediocrity," Parrish says. The answer for Datachecker was to spend $1.25 million and put 2,200 employees through a Philip Crosby Associates, Inc. training pro-

∽ EDUCATION IN U.S. INDUSTRY: A RACE TO CATCH UP

Human history becomes more and more a race between education and catastrophe.
—Herbert George Wells, 1920, from *The Outline of History*

Replace "human history" with "corporate survival" and Wells's words capture the essence of the struggle many technology companies face. The sorry story is that some industry leaders fear catastrophe is gaining the upper hand. They point to some disturbing evidence:

1. In the United States, it takes 10,000 high school sophomores to produce 20 science and engineering Ph.D's. Fewer and fewer students, however, are studying science and engineering at the high school level, says the National Science Foundation.
2. A study of 12th graders' math skills in 15 countries ranks U.S. students 14th in algebra and 12th in calculus and analytic geometry, according to the International Association for the Evaluation of Educational Achievement.
3. The same study compares science skills among 12th graders from 13 countries

who have at least 2 years of science education. The United States came in 9th in physics, 11th in chemistry, and dead last in biology.
4. By the year 2000, says the National Science Foundation, the U.S. demand for holders of bachelor-of-science degrees will exceed supply by more than 450,000.

To supplement inadequate public education, more and more technology companies are spending their own money to teach their employees. IBM, with $64 billion in annual sales, claims its yearly spending on internal education is greater than Harvard University's entire annual operating budget of $1 billion.

The education battle is being waged on a second front. Because many companies believe quality and literacy go hand in hand, a growing share of corporate education funding is being used to teach employees how to measure quality on the job and deliver it to the customer.

If the effort expands fast enough to overcome growing deficiencies in the U.S. system, education may win out over catastrophe after all.

gram. Crosby is a quality consulting firm owned by British-based Alexander Proudfoot Plc. "The timing was right, it really caught on," says Parrish. About halfway into the education process in 1987, National Semiconductor sold Datachecker to ICL of Britain for $126 million, at more than two times assets.

Whether the consultant is Crosby, Deming, Juran, Feigenbaum, or some other is immaterial. For the most part, the tools they teach are similar. What is key is that the company know why it needs the change and that top management is ready to personally join in taking on the task. "The key thing here is that this is not a program, it is a process, and that is why we start with senior management," says Howard Aaron, a quality consultant with Ernst & Young and an examiner for the Malcolm Baldrige National Quality Award. Likewise, Deming, the driving force behind statistical process control, will not consult with a company if he cannot have the undivided attention of top management.

It makes sense. To institute a strategic change, the strategists have to buy into the new concepts. As Aaron quips: "You're not going to go on a diet, you're going to change your eating habits." The easiest way to get top executives introduced to quality concepts is to send them to awareness seminars. One- or two-day sessions given by training and consulting firms serve to get the executives' feet wet in the concepts of continuous quality improvement and total quality.

A number of major companies suggest that top executives supplement awareness seminars with visits to other firms that have experience in quality. It could be a customer, a supplier, a competitor, or any company that has a similar corporate structure. Of great importance is that the company be willing to talk frankly about its experiences, says Rudell O'Neal, corporate quality training manager at Hewlett-Packard.

At the same time, other managers should be encouraged to get involved. This is usually an unstructured informational exercise that involves anything from attending courses to sharing reading lists. In the case of CalComp, Inc., of Anaheim, California., president William Conlin made *World Class Manufacturing*, a book by Richard Schonberger, required reading for all employees in the three manufacturing divisions.

CalComp believes the message is so important for U.S. industry that it has taken the unusual step of offering free to any CEO or top manufacturing executive a video and kit discussing how to implement world-class manufacturing. It has already sent dozens of them to companies large and small. Kits are available from David Schlotterbeck, the senior vice president who put together the company's world-class manufacturing program.

∾ Japan Can Teach Us Lessons

Many executives travel to Japan, the mecca of quality education, to get a first-hand look at state-of-the-art techniques. In the case of Texas Instruments, executives turned to an in-house division. In early 1986, executives from corporate headquarters traveled to TI's semiconductor fabrication facility in Hiji, Japan, to study how its total quality control process could be exported to U.S. operations. The year before, the Hiji plant won the Deming Prize, Japan's coveted industrial quality award, which dates to 1951. In addition, engineering teams were exchanged between the United States and Japan, and the company carefully observed their operating modes to gain insight into cultural differences.

Westinghouse Electric and Xerox, among others, made similar pilgrimages. In the early 1980s, David T. Kearns, then Xerox's chairman and chief executive, made numerous visits to affiliate Fuji Xerox, Ltd. in Japan, also a Deming winner. He returned with ideas for applying total quality throughout the corporation. In 1983, Kearns set in motion the company's Leadership Through Quality process, which helped the company win a Baldrige Award in 1989.

Westinghouse executives made similar trips in the late 1970s and early 1980s and took with them union leaders and technical experts, says Nathan Moore, manager of quality programs for Westinghouse's productivity and quality center. Westinghouse's Commercial Nuclear Fuel Division was a Baldrige winner in 1988.

When the top executives bring their new-found knowledge back inhouse, the immediate task is to set the agenda for changing the culture. That involves communicating the cultural changes that have to be made and educating and training the work force. For a number of leading firms, the starting point for that communication has been a policy statement made by the chief executive. For example, three years ago at Harris Corporation., John Hartley, the chairman, president and CEO, penned his company's mission statement. It reads: "Harris must be perceived as a company of the highest quality in every aspect of its business activity."

IBM believes it has refined the mission statement down to its essence, based on the philosophy of founder Thomas Watson Sr. The statement has three elements that all relate to quality:

- respect for the individual
- commitment to customer service
- pursuit of excellence

Once the work force believes that management is serious about change, the task is to take all the ideas and tools that have been accumulated internally and fit them into a cohesive framework. It is critical that all employees receive the same message. They may not all be given the same tools and techniques (manufacturing may not need tools learned in marketing and vice versa), but they should be speaking the same language.

∾ Developing the Curriculum

By all accounts, developing the in-house curriculum is the most critical step in the success of instilling a change in culture. The most successful companies pick and choose techniques and philosophies from as many sources as they can, rather than relying on one methodology. Synthesizing and customizing are the keys, according to Ken Wathen, manager of corporate quality advancement at Eaton Corporation and training board chairman of the American Society for Quality Control.

In-house education is important for two reasons: it creates a sense of ownership for the employees and it is flexible enough so that as the company's quality process matures, training and education can keep pace. As new theories and techniques come on the market—such as quality function deployment—they can be integrated into the curriculum.

Introduced to the United States from Japan in the late 1980s, quality function deployment is a process in which a company integrates customers' requirements into every phase of a product's development, from design and manufacturing to sales and marketing. It involves improving communication with the customer and building teams among departments in the company.

A feeling of ownership comes from designing the curriculum and choosing what is taught. It also comes from involving the employees in the education process. Sometimes that means letting design engineers teach a course on quality function deployment. It can also mean teaching while doing—studying a real problem and fixing it. Some companies insist that cross-functional work teams receive the same training together, because that is how they will work on the job.

San Diego–based Cipher Data, Inc., for one, is taking the in-house approach to education. This maker of tape and optical disk drives, a medium-sized firm with sales in the $200-million-a-year range, established a total quality college in 1988. One inducement was a grant of $1 million from the state of California's unemployment insurance fund, which sets aside money for training and retraining. The company's school even has its own dean.

So far, Cipher has put more than 800 of its employees through 40 hours of basic quality education and is gearing up to take the college to its Ann Arbor, Michigan, and Singapore operations. The courses cover process flow diagrams, process performance measurements, data collection and analysis, and statistical process control, says a spokesman.

Work teams at Harris' Corporation's Melbourne, Florida, semiconductor division, composed of on-line operators, engineers, and managers, used their training to make some significant changes. In one case, work-in-process inventory was reduced by 50%, cycle time dropped by 75%, and yield rose by up to 40%, says Sharon Sines, vice president for product assurance at the division and overseer of quality training. But Harris is not stopping there. The division has presented a plan to upper management to restructure the facility along functional lines and to revamp the incentive program to make striving for quality even more appealing to employees, she says.

Establishing a cross-functional position of vice president for quality is usually also necessary to help facilitate the education process. That individual should have direct reporting responsibility to the president and chief executive.

Companies tend to differ most when it comes to deciding what gets taught to whom. Some teach managers and line workers identical courses, reasoning that everyone should hear the same thing, according to Donald Hopkins, business manager at Corning Quality Systems, a division of Corning Inc. That goes beyond just awareness courses, he says, and should include skill training in statistical process control, problem solving skills, and communication.

Other companies are more selective. They might teach managers only the basics on the grounds that overloading with skills that are not going to be used is a waste of resources. For instance, general managers in manufacturing might get awareness training and instruction in group dynamics but would not be taught the details of statistical process control or design-of-experiment tools. Those would be given to such practitioners as manufacturing engineers.

Also, according to some experts, it makes little sense to blitz employees with everything they will need to know about quality for the next 10 years. "Avoid forced, mass training," suggests Jim Houlditch, senior vice president for quality, reliability, and operations at TI's Defense Systems and Electronics Group. The best results come from training that is given today and used today, he says.

~ The Payoff for Education and Training

Much of what a company teaches in quality-education programs depends on the funds available. While average spending by companies on in-house quality training is not available, a commonly quoted cost for awareness training runs at an average of $200 per employee. The return on even that basic investment is positive, companies say. While IBM does not tally its spending on training, says Robert Talbot, director of quality for the corporation, "We'd be making zero profits if not for robust design." A robust design is one that is highly tolerant of variations from specifications of individual components.

Some companies are more willing to divulge the savings they have achieved through training and education. For instance, at Harris's Semiconductor Division, the cost of quality dropped in 1988 by 15%. At TI's Defense Systems and Electronics Group, the division gets a return of $4 in improvements for $1 spent on training, says Houlditch. At Varian Associates, Inc., Edward Stone, director for corporate quality, puts the return on investment from quality training as high as 700% and claims to have saved $7.7 million in improved quality in the first 6 months of fiscal year 1989.

For some companies time is the critical element. Putting an entire company with thousands of employees through an education program can take months and even years. A financial crisis will not wait for the implementation council to come up with a master plan. In that case, spending big bucks to bring in an outside consultant to guide the process may be the only option. If a company choses that road, it should keep in mind a couple of things: it's best to hire a consultant with a track record and ask for and check references. Even so, using a consultant should only be considered a short-term catalyst in a continual process, experts say. An internal training program should be developed to ensure long-term continuous quality improvement.

A common thread for many corporations is to have some kind of train-the-trainer program in place whereby employees at different levels learn how to teach. In this way, the knowledge cascades through the organization and is controlled by those who use it, not by outsiders. This reinforces the idea of ownership.

At TI's Materials and Controls Group in Attleboro, Massachusetts, the visibility of top managers is a priority in training. All group presidents and vice presidents teach a 40-hour, 8-week course on Juran's quality improvement methodology at least twice, says Werner Schuele, vice president responsible for total quality control. The group has only two full-time, in-house trainers who focus on management development. It relies primarily on employees for this course and on outside consultants for tools training.

A bone of contention in corporate America is how much education should be given to new employees. Most companies offer at least an orientation program for new employees and in some cases provide much more of a grounding in the corporation's culture. is cited Every quality professional surveyed cites the lack of formal education in quality at the university level as a major stumbling block to U.S. competitiveness. While corporate-based education is a way to remedy that, many companies are beginning to work with universities and colleges in their areas in an attempt to improve the situation. Some companies, such as Motorola, teach math skills so their employees can work with Pareto charts and statistical process control, but in many cases employees also need help with remedial English.

While universities generally get a "D" in quality training, there are some exceptions. The Massachusetts Institute of Technology created a joint program with its Sloan School of Business and its engineering school to offer a graduate level program called Leaders for Manufacturing. The program includes some seminars on quality control and concurrent engineering and regularly brings in specialists from industry to interact with the students. Discussion is under way to expand the program, says Don Clausing, the Bernard M. Gordon Adjunct Professor of Engineering Innovation and Practice.

Other universities such as the University of North Texas, are setting up centers for productivity and quality as ways to interact with industry. These tend to be money-making consulting operations but at the same time serve to enhance the flow of information from industry to academia.

⟶ More Statistics for Engineering Students

Because of the importance of statistics in achieving quality, industry devotes huge amounts of money to training employees in statistical methods for quality improvement. Unfortunately, universities have given very little emphasis to statistical subjects. That situation might soon change, however, at least for engineering undergraduates.

The Accreditation Board for Engineering and Technology (ABET), the body that accredits university and college undergraduate engineering curricula, began in mid-1990 to review its policies regarding statistical education. With backing from the American Statistical Association, ABET has been developing plans to require more applied statistics in engineers' course loads. "What we are trying to do is find out where we are and where we should be aiming with statistics and stochastic methods within the engineering curriculum," says David Reyes-Guerra, executive director of ABET. A task force, with representatives from a few professional organizations, was formed to help write new guidelines on the role of statistics in engineering. At the top of the agenda was the question of what should be taught and by whom. In the past, at most, engineers currently got one or two courses in theoretical statistics, usually taught by the mathematics department. If students don't see a link between the subject and practical problems, the material usually goes in one ear and out the other. With a revised, more practical approach, it is hoped that engineering faculty will become more aware of the importance of statistics.

The real issue, says Craig Barrett, senior vice president at Intel Corporation's Chandler, Arizona, microcomputer and components group, is that the engineering faculty are the ones that need the education. Many are out of touch with what goes on outside their ivory towers. "Deans should get their faculty in line," he says. Based on some informal surveying, Barrett estimates that only about 20% of undergraduate engineering students from the top schools take statistics, compared with 80% for business gradsuates. Universities argue that applied statistics might be a good idea but that the course load for engineers is already heavy. Adding more courses is just not possible. The answer, suggests Jim Houlditch, senior vice president for quality, reliability, and operations at Texas Instruments' Defense Systems and Electronics Group, is to add some of the new concepts, including statistical methods and concurrent engineering, for example, to existing courses.

Reyes-Guerra suggests that an initial step would be to change the wording in ABET's criteria for teaching statistics to engineers from "should" to "must." Then more concrete recommendations will be developed.

Statistical methods are only a part of the new quality push, but they represent a foundation on which quality programs are built. In fact, statistical methods are so important they should be viewed as key assets valued on a par with intellectual property and trade secrets protected by the company, argues David Fluharty, chairman of the American Statistical Association's quality and productivity section.

Fluharty offers some recommendations aimed at helping managers better manage the use of statistics:

1. Understand what questions to ask when reviewing statistical applications.
2. Balance encouragement of initial uses of statistics with a need to improve the use of statistical technology.
3. Understand the need to integrate statistical, scientific and systems approaches.
4. Understand the relationship between statistics and other disciplines and accept the limitations of statistics.
5. Separate science from sales pitch both in the work of statistical consultants and in the content of presentations to employees.
6. Determine who in the organization needs to know what, and how to deliver sound, effective, and efficient training.
7. Insist on linking applications with statistical training.
8. Integrate statistics with managerial philosophy as well as with business, organizational, quality, and individual training plans.
9. Conduct scientific studies of what works in training and application.
10. Understand that the full benefit of the use of statistics in business can only be achieved in a culture that looks at data with objective skill.

Chapter 18 ~

How Companies Manage Education and Training

Three companies among many that give special emphasis to education in achieving a total quality culture are Intel Corporation, Rockwell Corporation's Digital Communications Division, and Hewlett-Packard Co. Here's how each of them tackled education issues.

~ Intel: Employees Go to School at Intel U.

In the early 1980s, Intel Corporation considered itself well ahead of the pack in quality education for its employees. As the nation's third-largest semiconductor maker (after Motorola and Texas Instruments), the chip and computer systems maker based in Santa Clara, California,had been upgrading its quality-control methods since its founding in 1968. Like many other companies, Intel had adopted the quality-circle approach during the late 1970s and moved into statistical process control in 1983. Employees even had their own private school—Intel University, founded in 1982—which offered highly tailored work-related courses at each of the company sites.

Then came the realization that even these extensive efforts fell far short of what it would take to remain globally competitive. The revelation began with some Intel executives, who happened to be studying the training of workers, during a series of visits to Japan during the early 1980s.

"At the time, the average assembly worker in the United States was receiving 4 or 5 days' training upon hiring," recalls Scott Pfotenhauer, total-quality marketing manager at Intel's Embedded Control and Memory Group (ECMG) in Chandler, Arizona, "and our company was running about 20% higher than that. In Japan, however, the average worker received 6 months' training." Although the training period included production work, by the end "the Japanese worker understood not only his own job and the other guy's job but also how all their jobs affect the whole process," according to Pfotenhauer.

Those visits led to some dramatic changes in the way Intel looked at employee education. For one thing, training for company workers now runs over periods of weeks or months instead of days, according to Evan Nielsen, total-quality strategic planner for the Chandler operation. For another, senior-level managers who were once excluded from training are now expected to sign on along with everyone else; even CEO Andy Grove has participated in off-site management programs. By 1989, says Pfotenhauer, Intel was spending about $40 million a year on worker training, including tuition reimbursement for outside schooling along with its own internal programs.

Intel's Chandler operation, a contender for the Baldrige Award, clearly hopes the investment will pay off in its overall quality ratings. Even though the unit failed to capture the award, the company says the focus on worker education netted some impressive returns:

- Average yield levels across the company's operations rose from 94% to 98-99%.
- The company's employee-empowerment process, begun in 1987, has so far saved Intel more than $17 million, according to David Hall, director of the quality technology group. In just one instance, a group of workers decided that the filters in a deionized-water unit could safely be changed once instead of twice a year, for an annual savings of $55,000.
- Where Intel typically used one operator per machine in its wafer fabrication, assembly, and test operations in 1989, each operator now runs as many as eight machines, as well as assuming responsibilities for routine maintenance of the machines. Because each worker is a genuine expert at his or her job, according to Hall, employee morale is significantly higher.

Much of the credit for these and other improvements goes to Intel University, which offers regular classes at five company sites in the United States: Santa Clara; Chandler; Albuquerque, New Mexico; Folsom, California; and Hillsboro, Oregon. The university originally offered courses pegged almost exclusively to corporate managers, says Richard Ward, manager of Intel University, with production workers trained on the plant floor. Now, he says, "all of the company's education functions are being brought under the university's auspices."

While many other in-house corporate training programs are geared primarily to boosting workers' on-the-job skills, Pfotenhauer explains that the aims of Intel U. are much broader. Every Intel employee, says Pfotenhauer, needs to be not only functionally competent—that is, in programming, fabrication, or some other technical skill—but also versed in the Intel culture and in total business skills. One course even offers tips for promoting directness in business and personal relations. A management review committee, which includes the director of human resources and the vice presidents of marketing and manufacturing, keeps the university's course content focused on the company's needs across all disciplines.

Intel's employee-education process goes far beyond the usual company indoctrination, although such courses may be part of the total educational program. The first step is the creation of a training and development plan, which involves custom-tailored, one-on-one sessions between the employee and her or his immediate supervisor. Except for some introductory courses aimed at certain new trainees, none of the classroom sessions is mandatory; still, "just about everyone signs up," says Hall.

Once a plan is laid out, the employee registers for the appropriate courses according to a schedule that is agreeable to both the employee and the supervisor. The courses vary in length, some running only half a day and others running several days;

〜 TODAY'S LESSONS: READING, WRITING, QUALITY

Many Silicon Valley technology companies are going back to school in order to prepare for the Malcolm Baldrige National Quality Award competition. Local companies, both large and small, are looking to San Jose State University's Center for Productivity and Manufacturing Engineering as a unique resource for objective quality education and training in quality methods.

Jay Pinson, the university's Dean of Engineering, established the center in 1983 and expanded the school of engineering facilities with a $45 million investment, including $15 million from industry. Since 1988, the center has been training local electronics executives and professionals in all aspects of quality.

Courses include everything from 1-day top management sessions to 3-day courses on performance measures, design of experiments, statistical process control, and total quality management. "Applications are our niche," says Bob McQueen, the center's director.

Corporate support is strong, claims McQueen. So far, the class roster has included the likes of General Electric Company, Unisys Corporation, Sun Microsystems, Inc., Apple Computer Corporation and Intel Corporation. Individuals instrumental in developing the program include Bob Puette, president of Apple USA, and Rick Willson, an Intel quality improvement professional.

While corporate support is needed, it is the smaller high-tech suppliers that face the most difficult challenges, says McQueen. They are the Achilles heel in the efforts to raise the general level of quality in American manufacturing. Smaller companies often lack such resources as training departments and are under increasing pressure from companies such as Motorola that require them to apply for the Baldrige Award.

Helping companies learn from the Baldrige Award process is a special focus at the center. "Sometimes, companies are in a situation where they can answer all the application questions but can't build results in their organization. Our model shows which systems within the organization are linked to the Baldrige requirements," says McQueen.

The model breaks down six corporate areas—strategic planning, operational controls, technical systems, human resources, education and training, and continuous improvement functions—so students can identify where they need to strengthen their internal controls. But that's only the start of the training. "Once you identify where you need to improve, you then need knowledge. We are knowledge providers," says McQueen.

For more information on courses, contact the center at (408) 924-4144.

---

the acquired information is immediately reinforced by on-the-job utilization. The subject matter is described briefly in the Intel University catalog (which is updated quarterly), along with the job titles and descriptions of potential enrollees. Under the general heading of marketing, for example, available courses include "Presenting for Results" and "Situational Leadership." The category of "Systems Manufacturing" lists courses in statistical process control, project management, and board identification. Unlike most courses, there are no passing or failing grades, according to Nielsen: "The main purpose is to transfer knowledge that can be put to work immediately."

Most of the courses are not taught by academicians but by experienced Intel employees—technicians, engineers, secretaries—who are fully familiar with the concepts and technologies. Qualified instructors now number between 450 and 500, according to Pfotenhauer. Other employees also can train to become instructors. On those occasions when Intel personnel lack the experience to teach a subject, the company brings in outside experts. In 1987, for example, outsiders initiated a course on conducting effective meetings; the course is now taught by Intel employees.

Intel concedes that, except for a few largely anecdotal examples, the company does not formally track the effectiveness of its employee-education program. Executives are convinced, however, that its training processes have resulted in enhanced problem-solving techniques, better experiment design, and more responsible and innovative employees. As a result, there is no sign that the company is planning to cut back on any of its education investment.

∼ Rockwell: The Aim Is Total Customer Satisfaction

When Gilbert Amelio was appointed president of Rockwell International Corporation's Digital Communications Division (DCD) in 1983, his major target was to improve customer satisfaction. Amelio (more recently chief executive for National Semiconductor Corporation) challenged the division to improve quality by a factor of 10 and to become at least twice as good as the nearest competitor. A major key to making that happen has been quality training and education for all 3,500 DCD employees.

Today, the division, based in Newport Beach, California is the leading supplier of facsimile modems in the world and sells the lion's share of its products in Japan, where most of the world's fax machines are manufactured. "Without the quality improvements, we would not be in the Japanese market right now," admits Lee Troxler, director of quality for the Digital Communications Division. Dean Niccarron, vice president of technology at In-Stat Inc., a Scottsdale, Arizona, market research firm, estimates that by 1989 Rockwell owned nearly 70% of the worldwide fax modem market, which topped $105 million in 1990.

Amelio later moved up the Rockwell corporate ladder, becoming president of Rockwell Communication Systems, but his legacy remains. Since 1982, DCD has managed to

- reduce the defect rate on the modems by a factor of 10 to below 100 parts per million
- bring down customer reject rates by 100%
- improve throughput time by approximately 400%
- achieve a 20-fold gain in reliability for the product, which allowed Rockwell to extend its warranty from 1 to 5 years
- improve gross profits even though product price has fallen by a factor of 36 since 1982

Rockwell's current quality program, called total customer satisfaction, purports to put the customer at the center of all employees' responsibilities and applies to both external and internal customers. Says Troxler: "We choose to pursue a particular effort only if it has a noticeable impact on the customer." Keeping the customer in mind helps the division establish its priorities. DCD is working to improve cycle-time measurements, customer returns, and services to customers. These efforts involve every function, including the financial and human resources departments.

Applying for the Malcolm Baldrige National Quality Award in 1990 helped the Digital Communications Division hone its long-standing quality program. While DCD was not a finalist, Troxler agrees the Baldrige application process enabled the division to audit many diverse programs and redefine them within the larger framework of a quality system.

The Digital Communications Division used the Baldrige application report as the cornerstone for a new quality training class for all DCD employees. Troxler estimates that it will take 12 months for all of the division's employees to take this class. "We'll use the Baldrige report to help our employees understand what their piece of the total-customer-satisfaction pie is," Troxler adds.

DCD has set a 5-year training schedule for all personnel associated with production operations, engineering, management, finance, personnel, and marketing. Each employee receives training specific to his or her function, as well as general quality training. Top managers are sent to off-site workshops conducted by outside consultants—16 of these management workshops have been held since 1984. In 1987, DCD established a quality council of senior executives that meets weekly to discuss quality improvements, monitor major quality indexes against goals, assign quality teams to resolve issues, establish defect-prevention actions, and guide strategic quality plans.

Bottom-up training has focused on statistical process control (SPC) methods for the engineering and production staff and, more recently, design of experiments (DOE) training. SPC training has been an ongoing effort at the division, but applying it has not been completely successful in every case. "In some cases, we found that we were applying SPC to the output rather than to the variables which were causing the operations to go out of control," says Troxler.

Through the use of DOE, however, employees are better able to identify and monitor the variables that may cause the line to go out of control before that ever happens. Says Troxler: "The object is not to stop the line when there's a problem—it's to find the problem and fix it long before that point."

Rockwell intends to apply DOE to all the divisions' manufacturing operations. The Digital Communications Division has been using DOE in its 5-inch wafer fabrication area and in the assembly area with some success. For instance, DOE has been applied in the assembly area to come up with an optimum temperature–power combination that provides a better yield for bonding. As a result, the rejection rates on bonding issues dropped by a factor of three over a 6-month period, according to Troxler.

Already 128 engineers have received formal DOE training—an average of 32 hours each, with 100 factory and design engineers slated for DOE training. The training is initially done through the use of outside experts, including Keki Bhote, a Motorola quality expert, and then brought in house.

The total cost of educating the engineers to date has been $40,000 in fees to consultants. Troxler says the savings have been measured at 10 times that amount. "The cost savings are so substantial—and such a given—that we're getting to the point where we may simply stop calculating them," he says.

On the factory floor, all operators must go through a skills-certification process that ensures that they understand their job and its requirements and that they have the proper skills to perform that job. The operators undergo a recertification process every 6 months.

Since 1984, a total of 1,142 wafer-fab employees have attended 32-hour classes on direct-labor training for a total of 36,544 hours. Another 154 employees have attended 24-hour classes on operator-training. In addition, 115 lead operators have each attended 40 to 80 hours of workshops on train-the-trainer techniques, for a total of 4,600 hours. These lead operators can then train other employees on the factory floor.

Having seen the tremendous quality improvements brought about by increased training in production areas, the division is making a bigger push into quality training for administrative and clerical employees. DCD plans to use training experts to instruct white-collar employees in cycle-time management and project-oriented quality-awareness techniques. For instance, the finance department is trying to improve the reaction time on its authorization requests system for capital equipment. Through cycle-time management training, the department hopes to process requests more quickly while reducing the number of errors. The secretarial staff is developing a training program for secretarial and clerical employees using input from top-rated secretaries to determine course content. The human resources department is trying to take a more proactive approach to personnel planning by developing a bank of qualified candidates for a position before a job opening actually occurs.

Beyond giving the employees specific quality skills, quality training has been a big morale booster, says Troxler. The training helps employees understand that every one of them can make a big difference. "People now see how they can improve a particular process or an operation where in the past they had only been frustrated bystanders," says Troxler. "There's a snowballing effect," he adds. "Everyone gets excited over this."

Through the annual employee attitude survey, employees of the Digital Communications division were asked to identify ways to improve quality and efficiency. Four of the top five recommendations dealt with employees:

- increase training and education
- improve teamwork
- increase employee responsibility
- improve trust between management and employees
- upgrade or purchase tools and equipment

Along with expanded training and education programs, the Rockwell division has designed a number of awards and programs to improve morale and stimulate quality improvement efforts from the bottom up. Following are a few of the programs:

- *Quality Recognition Award*—recognizes an individual or team that has contributed the most to the quality improvement effort and annually awards a plaque
- *Annual Employee Attitude Survey*—allows employees to anonymously share their evaluations of the division's operations, leadership, goals, and performance. There is a quality section within this survey
- *Employee Suggestion Program*—rewards employees with money if their suggestions save the division money. Employees receive 10% of the first year's cost savings up to $10,000. Since 1985 the program has saved the division more than $2 million and returned $144,000 to the employees

- *Coffee Klatch Program*—allows small groups of employees to speak with top management in an informal setting
- *Instant Compensation Program*—provides on-the-spot monetary recognition for specific employee or team accomplishment

∽ Hewlett-Packard: Focusing on Education from Within

Hewlett-Packard Co. (HP) has always had a reputation for selling high-quality computers, test-and-measurement instruments, and other equipment. For the most part, customers of the Palo Alto, California, company were willing to pay a premium for those products.

In 1980, however, the company found itself facing Japanese competitors selling comparably high-quality products at lower prices. HP had to change, says Rudell O'Neal, manager of corporate quality training and development. "We couldn't expect our customers to pay more than the competitor's price for similar products."

To become more competitive, HP in 1983 adopted total quality control, which O'Neal describes as "a management philosophy and operating methodology." The major tenets of Hewlett-Packard's TQC program are process improvement, customer satisfaction, and the use of quantitative tools.

Having settled on a methodology, HP was then faced with the question of how to train its employees to think quality. The company did not seek out high-priced consultants to solve its dilemma. Instead, it looked within. To start the process, in 1983 and 1984, HP combed its worldwide operations to learn what individual departments, divisions, and subsidiaries were doing in quality improvement training. The company adopted the programs that appeared to work well for a corporate quality training curriculum launched in 1985. Many of the courses came directly from HP's operations in Japan, Singapore, and Malaysia.

Quality has become a buzzword at many companies, notes Barry F. Willman, a computer analyst at Sanford C. Bernstein & Co. "What differentiates Hewlett-Packard from the rest is that they have institutionalized quality throughout the corporation." Worldwide, HP spends between $150 million and $200 million a year to educate its 92,000 employees, according to Neil M. Johnston, director of corporate education. When other costs are included, such as the time workers spend in classrooms and travel expenses, the total cost of education reaches $400 million to $500 million annually.

O'Neal says some 140 HP trainers spend more than 50% of their time running quality-education programs. Most, however, are assigned to specific business areas or departments and are experts in certain quality functions. One trainer teaches the employees software inspection, for example. O'Neal's five-member department oversees the decentralized quality-education operations.

For such an extensive program to develop and to succeed, according to O'Neal, it is vital that it be driven from the top. Two other principles are critical in successful quality training, say HP executives: The training must be closely linked to other corporate education programs, and top management must be involved. When HP began quality training in earnest in 1985, for example, TQC was taught to the company's work force in separate classes. Later, TQC education was integrated into orientation classes for new employees and training for new managers.

O'Neal believes that the visible involvement of management is the most critical element of implementing quality-improvement training and that any such effort delegated to a company's lower ranks is doomed. HP president John Young, for example, often kicks off employee orientations and other training sessions, describing the company's quality-improvement goals and how employees can help meet them.

HP now offers quality-improvement training for employees on three levels: general TQC education, followed by more job-specific quality training, and augmented by continuing education. General, or foundation, quality training given to all HP employees focuses on process improvement and understanding customer needs. O'Neal says this phase ensures that all employees share an understanding of TQC and process improvement, their advantages, how they fit into the HP culture, and how employees can use them. While much of this training is built into employee orientation, O'Neal says, the foundation training can total about 12 hours per employee.

"Everything is a process," says O'Neal. "Employees should be able to define their work in terms of a process." Employees are taught that they have the power to make or propose improvements in their process. Since they are performing the job, the thinking goes, they are the best judges of how the process can be improved.

Corporate education director Johnston provides an example from his own department, where an administrative assistant proposed improving the order-processing and shipping operations for educational materials. A team of three employees studied the process and recommended changes that cut late deliveries from 30% to 8%. The goal: 100% on-time deliveries. 'Thinking about quality gets embedded in the way you do your work," says Johnston. "People start thinking, "There must be a better way of doing this.'"

Foundation training makes all employees aware of quality-measurement methods, such as statistical process control and histograms. "That way we can read the meter on how we are doing," O'Neal says. Building on the foundation training, HP offers quality-improvement education targeting particular operations within the corporation. Managers, for example, are schooled in *hoshin kanri*, a methodology for planning and setting priorities developed at HP's Japanese subsidiary, Yokogawa Hewlett-Packard. *Hoshin kanri* teaches managers how to focus their attention on two or three critical issues facing the company or a particular department or division, in addition to carrying out day-to-day business chores. It also ensures that efforts up and down the corporate ladder are coordinated, with micro-goals supporting the overall objectives set by upper management.

The methodology helps management set a direction for dealing with key strategic issues (the words translate literally as "shiny metal" and "shows direction"— in other words, a compass). Top managers like chief operating officer Dean O. Morton use the methodology to set priorities and directions for HP for upcoming fiscal years, such as identifying the kinds of products the company will develop. Hewlett-Packard teaches the details of how to implement *hoshin kanri* at all levels of management, from the executive suite to shop-floor supervisors. But O'Neal says that all HP employees are made aware of its general principles.

"Quality maturity system" training offers managers a process for determining which quality-management tools and procedures should be standard operating procedures within a business unit, such as a division or a field office, and how close the

～ SOME TIPS ON QUALITY EDUCATION

New to the training game? Unsure how your company should start? Companies with years of experience in educating their workers offer some suggestions:

Varian Associates

Two issues are critical to the success of training programs, says Ed Stone, director for corporate quality. First, a company should be sure employees get a chance to apply the skills they learn to their jobs either during the course or immediately following. Second, when setting up internal training and education programs, managers should teach their own people.

Westinghouse Electric

Have a corporationwide philosophy but customize the program to local needs, says N.W. Moore, manager of quality programs for Westinghouse's productivity and quality center. Each division should be responsible for and have control over setting its own agenda based on its own unique needs. It should not be forced to accept a packaged program, he says.

Hewlett-Packard

It is more important to get top managers to visit other companies that have put quality initiatives in place than to hire smooth-talking consultants, says Rudell O'Neal, corporate quality training manager. Consultants tend to pitch one methodology, often ignoring issues that might be critical for your company, she says.

Harris Corporation

Educate employees on the concepts of internal customers, says Sharon Sines, vice pres-ident for product assurance for commercial products and worldwide manufacturing, semi-conductor division. If they can see the improvements that quality makes internally, they will have a better understanding of how it affects their external customers, she says.

Motorola

It is critical that the company have a common culture and that quality training starts at the top, according to George Fisher, president and chief executive officer. Senior executives have to demonstrate to the rest of the company that they are committed to the process of continuous improvement if they expect their subordinates to sign on.

Xerox Corporation

The goal is for all employees to speak a common language, says Sarah Turner, manager of planning and integration for corporate education and training. That means all employees should be given the same education in the concepts of total quality.

Corning, Inc.

All employees should go through the exact same quality training, agrees Donald Hopkins, business manager, Corning Quality Systems. It is extremely important that everyone in the company—all the way from top management to the factory floor—hear exactly the same thing, he says.

---

operation is to meeting that ideal. The concept grew out of HP's Singapore operations, where managers from individual departments reviewed each other's TQC progress.

Managers are also trained in "process of management," which examines the question of what makes an effective manager. The program uses exceptional managers, chosen by subordinates and superiors, as models for others to follow. Incorporated into the curriculum is the expertise of model managers, gained through in-depth interviews.

In the engineering arena, HP teaches quality function deployment, which focuses on improving the product definition and development process and teaching engineers how to design new products for manufacturability and reliability. Unlike the internally developed techniques HP teaches in most other training programs, the company borrowed many aspects of quality function deployment—concurrent engineering practices, for example—from automotive and other industries.

Programs like quality function deployment are critical for HP to meet its goal of a 50% reduction in break-even time for new products, according to a Sanford C. Bernstein report. While emphasizing time-to-market can incur "sacrifices in quality, features, or manufacturability," the report says, break-even time takes development costs into account, requiring gains in manufacturability and getting the product right the first time.

HP also emphasizes continuing education and keeping employees current in their fields of expertise. Johnston estimates that in the fast-moving high-technology arena, one-half of the knowledge possessed by an HP employee today will become obsolete within 4 years.

While some training is conducted in classroom settings, a great deal is done through consultation between employees and HP trainers, either in groups or one-on-one. O'Neal estimates quality training per employee can total 50 hours. "People at HP are very eager to learn," he says. "Just as we expect our processes to continuously improve, we expect our people to continuously improve in how they perform their jobs."

This is how O'Neal sums up the critical elements of HP's training programs:

- Focus is on process improvement and understanding customer needs.
- Programs provide employees with the tools to analyze and improve their job function and measure quality improvement progress.
- General quality improvement training is given to all employees followed by job-specific training.
- Quality training is largely on a consulting basis rather than in formal classroom settings.
- Emphasis is on employees understanding the need for quality improvement.
- Top management takes a visible role in quality training.

Chapter 19 ~

Views of the Gurus on America's Quality Crisis

Although the United States has embarked on a quest for total quality, there is still a long road to be traveled. No better sense of what lies ahead can be gained than by noting the views of some of the early seminal figures in the field. Four widely renowned gurus, who began preaching the gospel of quality long before the Baldrige Awards came on the scene, are W. Edwards Deming, Philip B. Crosby, Joseph M. Juran, and Armand V. Feigenbaum. Although they all tend to agree on some of the major principles of total quality, each of them also offers some unique perspectives that add deeper insights that can help guide future efforts.

An early success story in American business helps illustrate a major failing of a one-time view of quality in the United States when viewed in the light of the lessons these gurus have been teaching. It's the story of one of America's greatest retailers, Sears, Roebuck & Co. One reason consumers did so much business with Sears over the years was that no matter what a customer bought—lawnmower, back-to-school outfit, set of screwdrivers—the local Sears store cheerfully replaced it free of charge and without question if something went wrong. In the early 1980s, such a promise to repair or replace a defective piece of goods, which wasn't limited to Sears of course, was most Americans' idea of quality. Today, it is a symbol of everything that went wrong with U.S. manufacturers and service industries. It is a tacit endorsement of the misguided philosophy that says "doing it over can make up for any failure to do it right the first time." A replacement guarantee sounded pretty good as a way to keep customers in spite of occasional slip-ups "that were always bound to happen."

Now American managers are beginning to recognize how much that sort of credo really costs: The typical American factory has been spending up to 25% of its operating budget to find and fix its mistakes.

Also in the early 80s, Japan's attention to production process variables, so that the causes of problems could be identified and fixed, resulted in an average yield of about 50% for the memory chips most widely used in electronic equipment, according to one expert. That was in stark contrast to yields of only about 15% in the United States. That gave Japanese vendors three times as many chips to sell from the same production runs, greatly reducing costs. It also gave Japanese companies a deadly head start on advanced chip fabrication methods useful for expanding production to other

❧ WHAT IS THIS ELUSIVE THING CALLED QUALITY?

Defining *quality* is something like defining love. Most of us fall back on gut feelings, reasoning that we know when it's there and when it isn't.

As in matters of the heart, the so-called quality experts aren't much help; there is no single definition that satisfies everyone. Crosby offers what is probably the most succinct definition, calling it *conformance to requirements*, that is, the product should do what it is required by the user to do. By itself, of course, that defini-

tion says nothing about the effort and expense that may be required to assure conformance.

Juran addresses the issue by claiming that a quality product is "free from deficiencies" for the user and for the producer. Another gauge of quality, he says, is the cost to make it right, without excessive waste and rework.

Feigenbaum takes a more market-oriented stance: "Quality isn't specifications, or what advertisers or engineers say it is. It's what the buyer says it is."

---

types of integrated circuits. Japanese automakers also were reaping total quality benefits, needing only about half as many worker hours as U.S. plants to turn out similar models of cars and trucks, thanks largely to better design and higher quality parts.

The good news is that U.S. industry appears to have gotten the message. Although many companies are still crippled by the good-enough mentality, thousands of others have gotten the kind of religion that goes far deeper than the whirling dervish displays of the early 1980s, when loud pronouncements about quality were much more common than substantial commitment behind the scenes. Even some of the most demanding observers are beginning to see progress. "The overall quality of American companies has risen more dramatically than I've ever seen," says Joseph M. Juran, a veteran of more than 60 years of quality assurance campaigns. "The quality levels at some U.S. companies can match those of anyone in the world. Others are getting better, but still need a lot of work."

Lest Americans become too cocky, though, few think that the United States is out of its quality crisis. "I don't know many companies that regularly deliver defect-free products," complains Philip B. Crosby, another leading quality consultant, educator, and author. "In electronics, for example, companies are still testing and inspecting. They know basic quality principles, but many of them are trying to fix their own teeth."

To be fair, not all segments of U.S. industry hit the skids with respect to quality. In fact, the idea of total quality was an American invention, albeit one that, as so many others, was ultimately adopted and refined by Japan. Few countries can match America's performance in telecommunications, jet aircraft, agricultural equipment, and appliances. "Have you ever seen a Japanese refrigerator in an American home?" asks quality pioneer Armand V. Feigenbaum.

Clearly, though, something went wrong somewhere. In fact, something went wrong at several different levels. One level, explains Feigenbaum, was in the changing U.S. marketplace. For example, we tend to forget that Europe and Japan controlled less than 10% of the American auto market during the 1970s. By 1979, hundreds of thousands of Japanese cars sat unwanted on dealers' lots. That inventory disappeared almost overnight in the wake of the Iranian revolution and the ensuing oil crisis.

The average American buyer made some dramatic changes as well. "Ten years ago, we found that about 4 out of 10 U.S. buyers ranked quality at least as important as price in their buying decisions," says Feigenbaum. By last year, that proportion had doubled. "Customers today don't have time for waste or product failure," he insists. "By the 1990s, American companies will need to deliver essentially perfect products if they hope to compete."

The country's quality crisis goes much further than picky consumers, however. Across the manufacturing and service sectors, American companies, lulled, perhaps, by two competitively benign decades as Japan and Western Europe rebuilt their war-torn economies, adopted the good-enough standard. In some cases, those standards sounded impressive. For example, Crosby recalls that "several years ago, Western Electric had what seemed like a very rigid standard of allowing only one unsoldered joint per 100,000. But that said that you didn't have to do every joint right."

Unbeknownst to most Americans, Japan had set out in the early 1950s to solve what Juran recalls as a "quality crisis even worse than ours is today," thanks mostly to small producers of cheap ballpoint pens and transistor radios. Enlisting the help of American advisers such as Juran and W. Edwards Deming, Japanese upper managers quickly mastered the concepts of total quality control, then established massive worker education programs. As a result, today's Japanese companies do not routinely test products. Because of the zero-defect policy, according to Crosby, "things are done right the first time. The difference between them and us isn't just technique—it's policy."

Many factors obviously go into forging such a policy. Few are more important than the concept of process variation—how it arises, how it affects product yield, and how it can be controlled or eliminated. "We need much more knowledge on the subject of variation," says Deming, "whether it is among people or among products. Once we understand variation, we understand the losses that arise from tampering with a system."

Does Japanese industry understand variation? Apparently so, says Crosby: "In the U.S. semiconductor industry, many companies think that 20% or 40% yield is acceptable because it compares favorably with the competition. But Japanese producers get much higher yields by eliminating variables—they understand their processes and stick to them."

Even with the major improvement in overall quality during the past few years, most sources cite several fundamental weaknesses that still plague U.S. manufacturers. One point they all agree on is that top management at many companies has yet to prove that it is serious about catching up and keeping up in total quality. Too many executives still think they can delegate responsibility for achieving zero defects, says Crosby, "but it works only when the person at the very top lives and breathes quality."

Alas, more than one top manager has tried that approach, only to be shot out of the saddle for his or her trouble. "There are too many companies in which the chief executive makes a serious effort to discuss quality," says Feigenbaum, "and the board members sort of roll their eyes and wonder if they have a technician on their hands."

Another problem, according to Juran, is that quality is all too often planned by amateurs—typically designers and engineers who may be superb innovators but hardly authorities on how to instill total quality control throughout the company. Even if an engineer is skilled in the total-quality function, such a policy frequently sends the message throughout the company that quality assurance is the domain of one particular department.

By contrast, the most successful Japanese companies have sidestepped that pitfall via their company-wide education programs, which in effect make quality everyone's responsibility. "They teach the amateur to be a professional," explains Juran. "That eliminates turf battles and builds a lot of expertise into every individual."

Nor is the answer to total quality to be found solely in the potpourri of so-called management tools that have become so popular in the United States over the past few years. Deming, for one, insists that such tools—just-in-time, management-by-objectives, increased automation, granting more authority to employees, and even Crosby's famous demand for zero defects—do little more than help management duck its real responsibility. Deming tries to hammer home the point that management's charter is to become profoundly knowledgeable about people, processes, and the interactions between them. He recalls a company president who proudly noted that his employees were responsible for product quality. "But they aren't," scoffs Deming. "They can only try to do their jobs. Product quality is the president's responsibility."

Yet another fallacy is that quality can be assured through rigid testing at various points in the manufacturing process. "In the old days we tested the heck out of things," recalls Feigenbaum. "We'd test away 50% of their lifetimes before they reached the consumer." Moreover, testing is expensive and does not necessarily assure that all the tested components will function as a whole. Testing can rarely predict the lifetime of the finished product. Even worse, testing implies that defects exist in the first place, and so becomes a self-fulfilling prophecy. Quality-driven companies, in Japan and, increasingly, in the United States, boldly assume that defects don't exist. They back up that assumption with first-rate design and engineering, consistent manufacturing processes, and carefully planned vendor partnerships that guarantee that once a part is approved, it will never vary. Never.

America's business schools are another time-honored tradition that come in for bad grades from quality experts. Not only have schools not been setting the pace in quality assurance, they merely "perpetuate the present system of management," says Deming.

Many business schools, however, claim to be taking the proper steps to overcome past lapses in teaching total quality. But Feigenbaum cautions: "I'll take them seriously when I see full-time professors of quality. That acknowledges that quality isn't an ingredient of other courses. It's a technology, a teachable body of knowledge. But the only thing that's harder to move than a cemetery is the mind-set of a university faculty. I think that change will have to come from outside academe."

Not surprisingly, every quality guru has her or his pet remedy—usually served up at a relatively stiff price—for instilling excellence in a company. If you're lucky enough to get his attention, for example, Juran will consult with you for $10,000 a day. Alternatively, the Juran Institute, in Wilton, Connecticut, offers a variety of 5-day courses—for about $2,500 a head—including books and videos.

～ Taguchi Methods Couple Design with Production

Genichi Taguchi, for whom the Taguchi methods are named, is one of the top international quality gurus. He is a four-time winner of the prestigious Deming Prize in Japan. His specialty is product and process design that couples design with production.

Taguchi's design of experiments (DOE) focuses on building a cooperative relationship between design and manufacturing engineers. It combines statistical methods with engineering, enabling a company to achieve improvements in cost and quality by optimizing product design and manufacturing processes. The idea is to use only a few key measurements in order to find ways to improve a complex process with many interlinked variables. He also developed the quality loss function and a signal-to-noise ratio, measures that help engineers uncover problems early in product development and indicate whether low-cost improvements can be made.

From 1949 to 1961, Taguchi worked at Nippon Telephone and Telegraph (NTT) and was instrumental in implementing quality methodology for NTT's telephone exchanges. Even though Taguchi has been a consultant in Japan for more than 25 years, he only introduced his teaching and philosophy in the United States in the early 1980s. The first corporate clients included AT&T Bell Laboratories, Ford Motor Co., and Xerox Corporation.

While his teachings have been well received in the United States, Taguchi is not without his critics. The University of Wisconsin's statistics department, one of the leaders in applied statistics, is probably the most vocal: "Many of the techniques of statistical design and analysis Taguchi employs . . . are often inefficient and unnecessarily complicated and should be replaced or appropriately modified," reads one Wisconsin paper.

Philip Crosby Associates features popular and reportedly highly successful 2 1/2-day quality colleges for top executives and a 4-day course for managers, designed to teach executives how to manage for quality and instill the principles in every employee. Feigenbaum's General Systems Co., meanwhile, insists that it isn't enough to simply educate managers. There must also be a means of systematically implementing quality assurance procedures and policies throughout the entire organization.

Any company can do a great deal independently, however, to turn its quality performance around. The very first step, says Crosby, is to recognize that simply knowing the ABCs of quality control won't change a thing. "As long as a company is willing to set a policy of acceptable defects, nothing will change," he warns.

Once the right mentality is in place, adds Juran, it's critical to set up a corporate quality council, headed by the CEO, that sets specific goals and decides how to reach them. One way to eliminate production variables in the electronics industry, for example, is to form much closer ties between design and manufacturing. "The designer can't design whatever he or she wants," he says. According to Juran, the overall plan also must include a means of scoring and measuring yearly improvements in terms of defects per million parts, for example, or total customer complaints.

Instituting a total quality system will also call for some changes in how management looks at its workers, according to most sources. At the very least, the old adversarial relationships between workers and managers must go once and for all. Deming feels so strongly on the subject, in fact, that he puts joy in work at the very

~ PROFILES OF THE PIONEERS OF QUALITY IN THE UNITED STATES

Philip B. Crosby

When Philip B. Crosby was vice president and director for corporate quality for ITT in 1968, he waylaid the conglomerate's legendary CEO Harold S. Geneen at the elevators one day. En route to the 13th floor, Crosby gave what he now calls his elevator speech—a sermon on the cost of not doing things right at ITT, which Crosby put at some 20% of the corporation's annual sales. "That was when they started taking quality very seriously," he chuckles.

Crosby is now chairman of Philip Crosby Associates, Inc., a quality-management training and consulting company based in Winter Park, Florida, with offices in Great Britian, France, Singapore, Canada, and Australia. Founded in 1979, the company has thrived along with its so-called quality college, which defines quality as "conformance to requirements" and calls for a performance standard of zero defects. The college has trained hundreds of thousands of executives at client companies to become resident quality teachers. Crosby is the author of several best-selling books, including Quality Is Free and Quality Without Tears.

Crosby isn't lacking in critics. Although he is widely recognized as a dynamic and inspirational speaker, he's been called cocky and "a combination P.T. Barnum and the Pied Piper." If that bothers him, though, he is clearly comforted by the fact that his clients—including IBM, Chrysler Corporation, AT&T, and Digital Equipment Corporation—have made his consulting outfit one of the most phenomenally successful companies of its kind.

W. Edwards Deming

Top corporate executives can be excused if they hesitate to call W. Edwards Deming to the scene. Like it or not, the feisty octogenarian will brand them as morons, rip their management techniques to shreds (and those of many of his rivals), and huff that what they really need is joyous workers. Today's most cherished management strategies, he says, "squeeze from an individual his innate intrinsic motivation, self-esteem, and dignity and build into him fear, self-defense, and extrinsic motivation."

As one of the elder statesmen of the quality community, Deming can and does get away with all that and more. Now an independent consultant based in Washington, D.C., he became a legend for his work in Japan during the 1950s. In recognition of that work, Japanese manufacturers created a coveted national award for quality in 1951 and dubbed it the Deming Prize. Nine years later, he received the Second Order Medal of the Sacred Treasure from Emperor Hirohito.

Since then, Deming, who holds a doctorate in physics from Yale University, has consulted for manufacturing companies, railroads, hospitals, government agencies, and universities. Showing few signs of slowing down, he still travels worldwide, commands up to $10,000 a day for his services, and pulls thousands of eager disciples every year to his 4-day seminars.

A prolific author and musician, he is a member of the Science and Technology Hall of Fame and received the National Medal of Technology from President Reagan in 1987.

center of any quality program. "I estimate that only two managers in 100 take joy in their work," he says. "The other 98 are under stress from nonproductive tasks like battling takeovers and getting a good rating. They don't dare contribute innovation to their work."

Admittedly, the high technology industries pose some special challenges to the quality-conscious executive. One reason is the rapid pace of development in recent years. Whereas it took 20 years to bring television to market, for example, new semi-

Armand V. Feigenbaum

Unlike many of his colleagues and rivals, Armand V. Feigenbaum isn't much for fiery speeches or catchy slogans. A slim, dapper man in his late 60s, Feigenbaum, the president and chief executive of General Systems Co. in Pittsfield, Massachusetts, comes across more like the MIT engineer he is. But what he may lack in color and high visibility he more than makes up for in results. Before founding General Systems in 1968, Feigenbaum served as international director of manufacturing and quality with General Electric Co.

Since then, he's worked with some of the world's best known companies, including the Italian tire maker Pirelli, John Deere & Co. and Union Pacific Corporation, where his systems boosted rail traffic by 10% in a year. In 1988 he became the first American to receive the Georges Borel Prize, France's highest honor in quality control.

Feigenbaum generally shuns buzzwords, game plans, and easily memorized precepts. His message, however, may be summarized in a few basic principles. One is that many U.S. firms have a hidden plant—a work force equal to as much as 40% of capacity that exists solely to undo mistakes. Another is that quality isn't an abstraction but a technology that can be systemized and taught. In contrast to other quality superstars, Feigenbaum insists the key to quality is hands-on implementation (preferably custom-designed and installed by General Systems), not simply giving orders from an Olympian perch.

Feigenbaum is the author of Total Quality Control, one of the premier texts on quality improvement.

Joseph M. Juran

Ask Joseph M. Juran how Japan got to be the world's top dog in quality and he'll talk your leg off—and take a potshot at his own profession at the same time. "They trained the nation's entire corporate hierarchy in quality to become the best managers on earth," he says. "And they didn't do it with all these gurus that we have today, either."

Juran knows a thing or two about Japan. In addition to consulting to such companies as Motorola, Inc., E.I. du Pont de Nemours & Co., and Texas Instruments, Inc., he worked as a freelance adviser to Japanese industry during the 1950s, specializing in teaching managers how to find and eliminate the causes of poor quality. For that, he received the Order of the Sacred Treasure.

Now in his mid-80s, Juran is chairman emeritus of the Juran Institute in Wilton, Connecticut, a quality assurance educational and consulting firm he founded in 1979. Ever the footloose freelancer, he ended his daily involvement in the Institute in 1987, grumbling that "the thing had become my master."

Some of Juran's definitions lack the snappy clarity that distinguish those of others in the field. For example, he defines quality as "fitness for use," or "freedom from trouble." But he's still an avid evangelist of the quality gospel. "American companies know they have to improve," he says. "But they need more than slogans. Managers must take charge of their quality programs and tell their people how to do things better."

conductor technology is now commercialized in a matter of months. "That puts a special strain on the quality process at the front end," says Feigenbaum. "But rapid product development can also be an opportunity, especially in an industry that quickly integrates process and product. When you find a better way of doing things, you can usually put it to work right away."

The pioneers may differ on many of the finer points of quality assurance, but they agree on one thing: If American industry is to do more than just survive, the good-enough philosophy must be scrapped once and for all. It can't be done piecemeal or on a trial basis. It will demand a total transformation at every level of a company—design, engineering, assembly, marketing and worker and customer relations—and a wholesale repudiation of the management theory that Deming describes as "I win only if you lose." The result, promises Deming, will be "greater innovation in technologies and service, expanded markets, and higher material rewards for everyone. Under that system, everyone will win."

Resources
Recommendations for Further Study

It takes more than a few seminars, conferences, and lectures to be conversant in quality. For the serious and committed student, it takes a lot of reading and rereading, doing and redoing before the theory becomes a practical workable process.

But where to start? One way is to find out what some of the leaders in the field have been reading, and to get their recommendations. One respected industry practitioner is James Watson, a semiconductor group vice president at Texas Instruments, Inc., Dallas. Following is a selected sampling from Watson's extensive library of books, articles, and videotapes that he considers must reading. The 29-year TI veteran has his roots in operations and has been applying total quality control concepts to operations in the semiconductor group for several years.

Watson stresses that his list is not meant to be exhaustive. Instead, it should serve as a launching pad for further study by those hoping to gain a deeper understanding of the total quality process. He adds some pertinent comments on various sources that he has selected.

Some additional resources and comments have been added to Watson's list. These were contributed by Ron McCormick, also a vice president at Texas Instruments, and by Norman Rickard, Jr., vice president at Xerox Corporation's corporate quality office, and by the editors of *Electronic Business* magazine.

∼ *Watson's Books*

Albrecht, Karl. *At America's Service.* Homewood, Ill.: Dow Jones Irwin, 1988.
Albrecht, Karl, and Ron Zemke. *Service America.* Homewood, Ill.: Dow Jones Irwin, 1985.

Both are excellent books. They focus on servicing the customer: that is, all those things that go with the hardware that ultimately make up the product. This is where competitive advantage is made or missed. *At America's Service* is more of a how-to book while *Service America* is more anecdotal.

Brassard, Michael. *The Memory Jogger Plus.* Methuen, Mass.: GOAL/QPC, 1989.

Seven management and planning tools for continuous quality and productivity improvement are covered in detail. Specifically covered are: affinity diagram, interrelationship digraph, tree diagram, prioritization matrices, matrix diagram, PDPC chart, and activity network diagram. The book includes a basic introduction to the use and deployment of these tools.

Camp, Robert C. *Benchmarking: The Search for Industry Best Practices That Lead to Superior Performance.* Milwaukee, Wis.: Quality Press, ASQC, 1989.

Written by a Xerox employee, this book focuses on the benchmarking process as a means of seeking out the best practices that will lead to superior performance within an organization. Using a case history approach it details how to structure and conduct investigations, how to analyze and measure the opportunity for change, and how to implement an action plan to achieve improvement.

Davidow, William, and Bro Uttal. *Total Customer Service: The Ultimate Weapon.* New York: Harper Row Pubs. Inc., 1989.

This book presents a large number of anecdotes about improving service quality. It has particular relevance for the electronics industry, as it draws from Davidow's long experience in the industry both at Intel Corporation and at Hewlett-Packard Co.

Deming, W. Edwards. *Out of the Crisis.* Cambridge, Mass.: MIT, Center for Advanced Engineering Study, 1986.

Deming, W. Edwards. *Quality, Productivity, and Competitive Position.* Cambridge, Mass.: MIT Center for Advanced Engineering Study, 1982.

These two are total quality control classics by the quality guru who got the Japanese on the road to quality in the 1950s. Both books are worth reading, but in many ways Deming's philosophy is best taught by his interpreters, as they tend to explain the concepts of TQC more clearly. For example:

Scherkenbach, William. *The Deming Route to Quality and Productivity: Road Maps and Roadblocks.* Washington, D.C.: CeePress Books, George Washington University, 1986.

Ernst & Young Quality Improvement Consulting Group. *Total Quality—An Executive's Guide for the 1990s.* Homewood, Ill.: Dow Jones-Irwin, 1990.

This is a guidebook for top executives who want to radically redirect a corporate culture toward total quality. Topics include developing leadership commitment, measuring and rewarding performance, quality education requirements, employee involvement, JIT and quality, and building world-class suppliers.

Eureka, William E., and Nancy E. Ryan. *The Customer-Driven Company: Managerial Perspectives on QFD*. Dearborn, Mich.: ASI Press, 1988.

Written by an experienced American industrial engineer, the book shows how quality function deployment can be incorporated into any company, regardless of product line, service, or business environment. The book will answer the most frequently asked questions about QFD.

Feigenbaum, Armand V. *Total Quality Control*. New York: McGraw-Hill Book Co., 1983.

Feigenbaum's book, first published in 1951, is a classic. The main point of the book is that there is such a thing as total quality that extends across all segments of the business. However, his approach suffers because he believes the quality department should be responsible for quality, rather than top management adopting the quality methodology for themselves. It is written more for the quality professional than for the general manager.

Ishikawa, Kaoru. *Guide to Quality Control*. Lanham, Md.: UNIPUB, 1976.

Ishikawa's intention was to write a book for the quality circle supervisor that discussed the seven tools of TQC. As it turned out, it is just as appropriate for general consumption as it is for the specialist.

Ishikawa, Kaoru. *What Is Total Quality Control, the Japanese Way*. Englewood Cliffs, N.J.: ASQC Quality Press, Prentice Hall, Inc., 1985.

The "Old Testament" on quality control was written by Japan's foremost quality guru, who died in 1989. It is not the best book for learning techniques, but it offers useful insight and perspective on TQC because of the author's depth of experience.

Juran, Joseph M. *Managerial Breakthrough—A New Concept of the Manager's Job*. New York: McGraw-Hill Book Co., 1964.

This is Juran's classic. It is hard to fault him on anything he says. The concepts are clear and consistent with Japanese style TQC but are stated in Western terms. As with all the gurus, though, the biggest challenge with Juran is understanding his terminology.

Juran, Joseph M., and Frank M. Gryna, Jr. *Quality Planning and Analysis: From Product Development Through Use*. New York: McGraw-Hill Book Co., 1980.

This is a textbook about the quality of products and services needed by society. It provides insights into motivation, safety and liability, quality costs, information systems for quality, and quality assurance. The book includes real problems to provide the readers with a real-world perspective vs. traditional classroom theory.

Karatsu, Hajime, and Tokoki Ikeda. *Mastering the Tools of QC*. Tokyo; PHP Institute, Inc.

The book carefully explains what companywide TQC is and what the seven tools of quality control are. It explains all concepts in a very simple and direct approach.

Kume, Hitoshi. *Statistical Methods for Quality Improvement.* The Association for Overseas Technical Scholarship.

Statistical methods are effective tools for improving the production process and reducing defects. This book shows how to apply statistical methods to real world problems. It is useful both to beginners with little knowledge of statistical analysis and to experienced practitioners.

Lele, Milind M., and Jagdish N. Sheth. *The Customer is Key: Gaining an Unbeatable Advantage Through Customer Satisfaction.* New York: John Wiley & Sons, 1987.

This book's message is simple: Customer satisfaction is key to long-term profitability, and keeping the customer happy is everybody's business. If customers are happy with the value delivered by the company's products, if they feel they are valued and treated fairly, they will stay loyal to the company for a long time.

Mansir, Brian E., and Nicholas R. Schacht. *An Introduction to the Continuous Improvement Process: Principles and Practices* and *Total Quality Management: A Guide to Implementation.* Bethesda, Md.: Logistics Management Institute.

The Logistics Management Institute produced these two books for Pentagon contractors just beginning the quality process. They concisely sum up many ideas from a wide range of Japanese as well as American quality experts and also contain many examples and extensive checklists.

Mizuno, Shigero. *Company Wide Total Quality Control.* White Plains, N.Y.: UNIPUB-Kraus International, 1987.

One of the best books around on Japanese style TQC. If the choice had to be made, this book should be the one to read, even before Ishikawa. It contains one of the most thorough discussions of the topic; very readable.

Nemoto, Masao. *Total Quality Control for Management—Strategies and Techniques from Toyota and Toyota Gosei.* Englewood Cliffs, N.J.: Prentice Hall, Inc., 1987.

An interesting book written for managers that face issues similar to those of the automaker Toyota. Basically a good management philosophy book presented in a series of case studies drawn from the Toyota's history.

Scherkenbach, William W. *The Deming Route to Quality and Productivity: Road Maps and Roadblocks.* Washington, D.C.: CEEPress Books, George Washington University, 1986.

Reviews the Deming philosophy and how it is being implemented by business and industry worldwide. It is an introduction to and the analysis of the 14 fundamental teachings of Deming, which are the bedrock of the Deming approach. It is good, old-fashioned common sense on winning in today's marketplace.

Shores, Richard A. *Survival of the Fittest.* Milwaukee, Wis.: ASQC Quality Press, 1988.

A very interesting read and, like Mizuno's book, one of the best on the topic of TQC. Shores draws heavily from his management experiences at Hewlett-Packard Co. in describing how a company should position itself for survival.

Wheeler, Donald J., and David S. Chambers. *Understanding Statistical Process Control.* Knoxville, Tenn.: Statistical Process Controls, Inc., 1986.

An easily understood review and application for the industrial practitioner who may have had little or no formal training in statistical techniques. Extensive use of graphs, examples, and case histories facilitate explanations of how the techniques work.

～ *Other Books*

Cottle, David W. *Client-Centered Service: How to Keep Them Coming Back for More.* New York: Wiley, 1990.

Crosby, Philip B. *Quality Is Free: The Art of Making Quality Free.* New York: McGraw-Hill Book Co., 1979.

Crosby, Philip B. *Quality Without Tears: The Art of Hassle-free Management.* New York: McGraw-Hill Book Co., 1984.

Crosby, Philip B. *Let's Talk Quality: 96 Questions You Always Wanted to Ask Phil Crosby.* New York: McGraw-Hill Book Co., 1989.,

Dertouzos, Michael, Richard K. Lester, and Robert M. Solow. *Made in America.* Cambridge, Mass.: MIT Press, 1989.

Deming, W. Edwards. *Japanese Methods for Productivity and Quality.* Washington D.C.: George Washington University, 1981.

Fukuda, Ryuji. *Managerial Engineering: Techniques for Improving Quality and Productivity in the Workplace.*

Johnson, Perry L. *Keeping Score: Strategies and Tactics for Winning the Quality War.* New York: Harper and Row, 1989.

Juran, Joseph M. *Quality Control Handbook.* New York: McGraw-Hill Bood Co., 1979.

Juran, Joseph M. *Juran On Leadership for Quality: An Executive Handbook.* New York: The Free Press, 1989.

Kelada, Joseph. *Integral Quality Management: The Path to Total Quality.* Quafec, Inc., 1989.

Monden, Yashuhiro. *Toyota Production System.* Norcross, Ga.: Institute of Industrial Engineers, 1982.

Owen, D. B. *Beating Your Competition Through Quality.* New York: Marcel Dekker, Inc., 1989.

Peters, Thomas J., and Robert H. Waterman, Jr. *In Search of Excellence: Lessons from America's Best Run Companies.* New York: Harper and Row, 1982.

Rosander, A. C. *Applications of Quality Control in the Service Industries.* New York: Marcel Dekker, Inc., 1985.

Savage, Charles. *Fifth Generation Management.* Bedford, Mass.: Digital Press, 1990.

Schonberger, Richard J. *World Class Manufacturing—The Lessons of Simplicity Applied.* New York: The Free Press, 1986.

Shingo, Shigeo. *A Revolution in Manufacturing: The SMED System.* Stamford, Conn.: Productivity Press, 1985.

Shingo, Shigeo. *Study of the Toyota Production System from an Industrial Engineering Viewpoint.* Stamford, Conn.: Productivity Press, 1987.

Sontag, Harvey. *Corporate Perceptions: A Quality Primer.* Milwaukee, Wis.: ASQC Quality Press, 1989.

Thomas, Philip R. *Competitiveness Through Total Cycle Time.* New York: McGraw-Hill Book Co., 1989.

Western Electric. *Statistical Quality Control Handbook.* Easton, Pa.: Mack Publishing Co., 1977.

Wholey, Joseph S. *Operational Excellence: Stimulating Quality and Communicating Values.* Lexington Books, 1987.

⌒ Articles and Publications

Many of the entries in this section come from Quality Progress magazine, published by the American Society for Quality Control, Milwaukee, Wisconsin. *Quality Progress* is the premier monthly magazine devoted to issues in quality. It is in sync with total quality control as a management methodology, more so than most other periodicals in the quality world.

Beer, Michael C. Corporate change and quality. *Quality Progress,* Feb. 1988.

Bertrand, Kate. Marketers discover what quality really means. *Business Marketing,* April 1987.

Bhote, Keri. America's quality health diagnosis: Strong heart, weak head. *Management Review,* May 1989.

Bower, Joseph L., and Thomas Hout. Fast-cycle capability for competitive power. *Harvard Business Review,* Nov.–Dec. 1988.

Camp, Robert. Benchmarking: The search for the best practices that lead to superior performance. Parts 1–5. *Quality Progress,* Jan. through May 1989.

Cohen, Louis. Quality function deployment: An application perspective from Digital Equipment Corp. *National Productivity Review,* Summer 1988.

Desatnick, Robert L. Long live the king. *Quality Progress,* April 1989.

Drucker, Peter. The coming of the new organization. *Harvard Business Review,* Jan.–Feb. 1988.

Dumaine, Brian. A humble hero drives Ford to the top. *Fortune,* Jan. 4, 1988.

Dumaine, Brian. How managers can succeed through speed. *Fortune,* Feb. 13, 1989.

Finkelman, Daniel. If the customer has an itch, scratch it. *The New York Times,* May 14, 1989.

Gale, Bradley. How quality drives market share. *The Quality Review,* Summer 1987.

Gale, Bradley T., and Robert D. Buzzell. Market perceived quality: Key strategic concept. *Planning Review, International Society for Planning and Strategic Management,* Oxford, Ohio, March–April 1989.

Garvin, David. Quality on the line. *Harvard Business Review,* Sept.–Oct. 1983.

Goodman, John, Arlene Malech, and Theodore Marra. I can't get no satisfaction. *The Quality Review,* Winter 1987.

Guaspari, John. You want to buy-in to quality? Then you've got to sell it. *Management Review,* Jan. 1988.

Hammond, Joshua. Claude I. Taylor—Quality is the measure of value. *Quality Progress,* Aug. 1986.

Hauser, John, and Don Clausing. The house of quality. *Harvard Business Review,* May–June 1988.

Henkoff, Ronald. What Motorola learns from Japan. *Fortune,* April 24, 1989.

Kane, Edward J. IBM's quality focus on the business process. *Quality Progress,* April 1986.

Kearns, David T. A corporate response. *Quality Progress,* Feb. 1988.

King, Carol. A framework for a service quality assurance system. *Quality Progress,* Sept. 1987.

Persico, John, Jr. Team up for quality improvement. *Quality Progress,* Jan. 1989.

Quality control: An international concept? From "The evolution of Japanese management," *Japan Economic Journal,* Want Publishing Co., Washington D.C., Jan. 14, 1989.

Scholtes, Peter. Joiner—An elaboration of Deming's teachings on performance appraisal. Joiner Associates, Inc., 1987.

Scholtes, Peter R., and Heero Hacquebord. Beginning the quality transformation, part I; and Six strategies for beginning the quality transformation, part II. *Quality Progress,* July–Aug. 1988.

Sellers, Patricia. Getting customers to love you. *Fortune,* March 13, 1989.

Shapiro, Benson P. What the hell is market oriented? *Harvard Business Review,* Nov.–Dec. 1988.

Shapiro, Benson, et al. Manage customers for profits (not just for sales). *Harvard Business Review,* Sept.–Oct. 1987.

Shores, Dick. TQC: Science, not witchcraft. *Quality Progress,* April 1989.

Stratton, Brad. Payment in kind. *Quality Progress,* April 1989.

Sullivan, Edward. Douglas D. Danforth—A common commitment to total quality. *Quality Progress,* April 1986.

Sullivan, Edward. OPTIM: Linking cost, time, and quality. *Quality Progress,* April 1986.

Sullivan, L. P. Quality function deployment. *Quality Progress,* June 1986.

Tucker, Frances Gaither, Seymour Zivan, and Robert Camp. How to measure yourself against the best. *Harvard Business Review,* Jan.–Feb. 1987.

Waite, Charles L., Jr. Timing is everything *Quality Progress,* April 1989.

Electronic Business, Special Quality Issue, Oct. 7, 1991; *Electronic Business,* Special Issue: Commitment to Quality, Oct. 15, 1990; *Electronic Business,* Quest for quality—Special Issue, Oct. 16, 1989. Reprints of the editorial sections from these special issues are available from Cahners Reprint Service, Cahners Plaza, 1350 E. Touhy Ave., P.O. Box 5080, Des Plaines, Ill. 60018. (708) 635-8800.

The quality freeway. *Quality Progress,* July 1990.

What is leadership? What differentiates the most customer-focused companies? *Fortune,* June 4, 1990.

Is the customer always right? *Incentive,* May 1990.

Main, Jeremy. How to win the Baldrige Award. *Fortune,* April 23, 1990.

Quinn, James Brian, Thomas L. Doorley, and Penny C. Paquette. Beyond products: services-based strategy. *Harvard Business Review,* Mar.–Apr. 1990.

Dumaine, Brian. Creating a new company culture. *Fortune,* Jan. 15, 1990.

Reynolds, Larry. Promoting quality in the public and private sectors. *Management Review,* May 1989.

Hart, Christopher W. L. The power of unconditional service guarantees. *Harvard Business Review,* Reprint #88405.

Kanter, Rosabeth Moss. The new managerial work. *Harvard Business Review,* Nov.–Dec. 1989.

Stowell, Daniel M. Quality in the marketing process. *Quality Progress,* Oct. 1989.

Kearns, David T. Chasing a moving target. *Quality Progress,* Oct. 1989.

Making total quality happen: The Conference Board, *Research Report* #937, 1990.

Total quality performance: The Conference Board, *Research Report* #909, 1988.

Kacker, Raghu N. Quality planning for service industries. *Quality Progress,* Aug. 1988.

U.S. electronics in Japan? Not without that quality, Electronic Business, Aug. 1, 1988.

Quality awards: The OEM way of saying thanks, Electronic Business, March 15, 1988.

Can a company spend too much on quality? Electronic Business, May 1, 1988.

Inside Xerox: Moving in a quality direction. *Electronic Business,* April 1, 1987.

Quality quest: The revitalization of high-tech manufacturing, *Electronic Business,* Jan. 15, 1987.

Farrow, John. Quality Audits: An invitation to managers. *Quality Progress,* Jan. 1987.

Schonberger, Richard J., and James P. Gilbert. Just-in-time purchasing: A challenge for U.S. industry. *California Management Review,* University of California at Berkeley School of Business Administration, Berkeley, Calif., Fall 1983.

⮀ *Videotapes*

Deming Library series of videotapes. Chicago: Films, Inc. (800-323-4222).

An excellent overview discussion of TQC. Donald Petersen, chairman of Ford Motor Co., is featured very effectively talking about Ford's experience. Of use mainly as inspiration and as a motivator for managers who can relate to Petersen's predicament. Not designed to teach techniques or tools.

Enclyclopedia Britannica, tapes. (800-554-9862, ext. 6513).

Two 1988 videotapes that highlight TQC at Hewlett-Packard Co. One discusses TQC in a nonmanufacturing customer service area, the other talks about TQC in manufacturing.

National Quality Forum videos can be purchased from the American Society for Quality Control of Milwaukee, Wisconsin (414-272-8575). These are videos of the forums in 1987, 1988, and 1989. Texas Instruments, Inc. has used these tapes for education and motivation as part of its corporate world training service.

"If Japan Can, Why Can't We?" NBC White Paper, tape, 75 minutes, aired June 24, 1980. Chicago: Films, Inc. (800-323-4222).

Investigates the lack of American productivity growth and compares it with Japan's productivity efforts.

"Quality is Free" Philip Crosby, tape, 23 minutes. Winter Park, Fla.: Philip Crosby Associates (407-645-1733).

~ Conferences

The National Quality Forum
This annual forum is jointly sponsored by industry, *Fortune* magazine and the American Society for Quality Control (ASQC). It is one of the premier opportunities to hear top managers endorse quality as a management methodology. For information, contact the ASQC's Conference Department, 800-451-7557.

Total Quality Performance Conference
European Quality Performance Conference
Both conferences are sponsored by The Conference Board, a New York City–based executive board that examines relevant issues pertaining to management. Both are excellent venues to hear top managers discuss quality issues.

Annual Quality Congress
Sponsored by the ASQC, this conference is almost too big and tends to be unfocused, at least as far as TQC management is concerned. A lot of juggling of schedules has to be done in order to attend the most important sessions. For information contact the ASQC's Conference Department (800-451-7557).

The Quest for Excellence Executive Conference
Presented by the Department of Commerce in cooperation with each year's Baldrige Award winners, this conference has been held in Washington, D.C., in February. Contact the Department of Commerce or NIST for specific details.

ODI, a Burlington, Massachusetts, management consulting and training firm, offers techniques, instruction, and seminars for those seeking to establish quality programs. The firm assists companies and governmental entities to improve quality, increase productivity, and manage human resources. For information call 800-ODI-INFO outside of Massachusetts and 617-272-8040 in Massachusetts.

Glossary of Common Quality Terminology

The quality field may be the biggest kettle of alphabet soup outside the Pentagon. To help prevent a bad case of "acronymia," we've tried to keep the alphabetese and specialized terminology to a minimum in this book. But as anyone who has launched a quality program will testify, the special shorthand of quality is so pervasive, knowledge of at least a few special terms and abbreviations is essential to understanding the literature in the field.

Those readers who search further will find a lexicon rich in three-letter abbreviations and such code names as Taguchi methods and *hoshin kanri*. They'll find that *quality control* is a far cry from *total quality management*. To cut some of the fog, we offer a guide to a few of the common terms in simple language.

Acceptable quality level (AQL) Maximum acceptable percentage of defective units within a set batch size.

Computer-integrated manufacturing (CIM) Approach to production in which a hierarchy of networked computers control and monitor every step of the process, determining which parts, materials, and subassemblies will move to which stations according to which schedules, selecting tools and operations to be performed, controlling operations, and testing to ensure that prespecified tolerances are met.

Continuous flow manufacturing Method of organizing production so that parts/materials come in one end and finished assemblies or subassemblies come out the other. This may require duplicating some production equipment for each separate product line.

Continuous improvement process (CIP) Approach to steadily improving quality through identifying and correcting root sources of defects. These may in some cases be subtle factors with small but cumulative effects. To meet ever-tougher goals, it is important to examine as many ideas as possible, using both a collection of small improvements and major shifts in how things are done. This leads to benchmarking processes against those used elsewhere, including in other types of industries and businesses, and enlisting the best thinking of everyone possible.

Control charts Charts plotted in different ways (histograms, Pareto charts, scatter diagrams, etc.) used for observing manufacturing processes and to help reveal underlying patterns such as relationships between variables. These data are useful in tracking sources of variation so that they can be controlled or eliminated.

Fishbone diagrams Plots that resemble the skeleton of a fish, used to track back to the root causes of variation or problems. Causes are determined by constantly asking why something occurs and then designating findings along a cascading sequence of causes that can then be plotted in the fish-bone format.

Hoshin kanri Approach to policy deployment starting with broad, long-term strategic objectives to serve as guidelines, as teams throughout the organization develop more short-range, detailed targets for their own groups. The planning process follows a cycle introduced in Japan by quality consultant W. Edwards Deming.

Just-in-time (JIT) manufacturing Production system in which materials or parts are delivered as they are needed for assembly, rather than being kept in inventories or safety stocks.

Kaizen Continuous improvement approach to manufacturing adopted in Japan based on the teachings of Deming. Statistical methods are used to determine the sources of variation in a process, to improve the average output, and to eliminate root causes of variation. The result sought is zero defects, which, Deming pointed out, would increase customer satisfaction while at the same time boosting profitability.

Kanban A *pull* method of manufacturing, in which materials or parts are delivered from a preceding stage or station upon request (sometimes by sending a *kanban card*), rather than a *push* system in which work is done at each station as materials or parts arrive and then are kept as in-process inventory at each succeeding station. Inventory is thus kept low.

Materials requirements planning (MRP) Approach to inventory management linking bills of materials (from design) to procurement, based on production needs. For example, 2-year projections might be updated monthly or more often. Suppliers may also be included in the procurement loop, so they can plan their own production capacity and scheduling.

Mean time between failures (MBTF) Measure of the reliability of a product. It is a statistical average of the operating time between failures over a population of devices or systems, or a designated sample of the total population.

Pareto charts Bar graphs arranging the sources of process variation in descending order according to their influence, thus indicating which factors offer the greatest potential for reducing variation. Causes of failures can similarly be plotted so that they can be remedied in the order of occurrence.

Poka-yoke Foolproof system of design espoused by Shigeo Shingo, a consultant to major Japanese manufacturing firms. Equipment designed using this approach, for example, makes it impossible for an operator to push the wrong button.

Quality circles Teams that may include workers, supervisors, managers, and engineers, often from different functions or departments, that meet to discuss ideas for making improvements in manufacturing or other processes.

Quality control (QC) Process of ensuring product quality.

Quality function deployment (QFD)   Approach to design focusing on meeting actual customer needs rather than simply manufacturing to a set of predefined specifications. Matrix charts can aid the process by matching carefully defined customer requirements, determined by in-depth exchanges with the potential customer or customers, and then matching these needs to available resources. An organization can apply the serving-the-customer objective to all its operations, and define anything that does not contribute to this ultimate goal as waste, which needs to be minimized. Thus, such items as inventory, testing, paperwork for internal purposes, design changes, and rework of faulty units can all be viewed as waste, since they do not directly meet needs of the customer.

Robust design   Approach to design allowing considerable variation in parameters of components without degrading performance. Combining robust design with methods to reduce process variation can dramatically reduce defects in manufacturing.

Single minute exchange of dies (SMED)   Method devised for Toyota with the help of Shigeo Shingo to cut the time required to change dies for making automobile bodies from hours down to a minute, greatly shortening production time and increasing flexibility.

Statistical process control (SPC)   Tracking of variations in manufacturing processes, sometimes with the understanding that the line will be shut down if upper or lower limits of variation are exceeded. The data produced are used to help identify sources of variation so they may be reduced.

Statistical quality control (SQC)   Tracking of defects in parts and materials and defining pass or fail limits based on statistical measures.

Taguchi methods   Approaches developed by Genichi Taguchi, a Japanese engineer and management consultant, for optimizing the design of processes and products. Taguchi developed a *loss function*, determining the costs of variations from target values. His methods aim to reduce variation and find near-optimum operating levels with just a few designed experiments, even though many variables may be involved and relationships may be nonlinear.

Total quality management (TQM)   Broad approach to quality, including product quality but extending well beyond to virtually everything done by an organization for external as well as internal customers within the same organization (what marketing does for manufacturing, for example). Continuous improvement is sought toward measurable, ever more difficult quality targets.

Index

just-in-time (JIT) manufacturing, 42
Leadership Through Quality program, 40
management communication, 43
product design, 41–42
worker education and training, 189

Yogawa Hewlett-Packard. *See* Hewlett-Packard (HP)
Young, John (Hewlett-Packard Co.), 106, 188

Zytec Corporation, 92, 101–104
Deming principles, 101–104
just-in-time (JIT) manufacturing, 102
management by planning (MBP), 103
teamwork, 103
worker education and training, 103
Xerox Corporation and, 101